T0312878

Power Efficiency in Broadband Wireless Communications

OTHER COMMUNICATIONS BOOKS FROM AUERBACH

Anonymous Communication Networks: Protecting Privacy on the Web
Kun Peng
ISBN 978-1-4398-8157-6

Case Studies in System of Systems, Enterprise Systems, and Complex Systems Engineering
Alex Gorod, Brian E. White, Vernon Ireland, S. Jimmy Gandhi, and Brian Sauser (Editors)
ISBN 978-1-4665-0239-0

Cyber-Physical Systems: Integrated Computing and Engineering Design
Fei Hu
ISBN 978-1-4665-7700-8

Evolutionary Dynamics of Complex Communications Networks
Vasileios Karyotis, Eleni Stai, and Symeon Papavassiliou
ISBN 978-1-4665-1840-7

Fading and Interference Mitigation in Wireless Communications
Stefan Panic, Mihajlo Stefanovic, Jelena Anastasov, and Petar Spalevic
ISBN 978-1-4665-0841-5

Green Networking and Communications: ICT for Sustainability
Shafiullah Khan and Jaime Lloret Mauri (Editors)
ISBN 978-1-4665-6874-7

Intrusion Detection in Wireless Ad-Hoc Networks
Nabendu Chaki and Rituparna Chaki (Editors)
ISBN 978-1-4665-1565-9

Intrusion Detection Networks: A Key to Collaborative Security
Carol Fung and Raouf Boutaba
ISBN 978-1-4665-6412-1

Machine-to-Machine Communications: Architectures, Technology, Standards, and Applications
Vojislav B. Mišić and Jelena Mišić (Editors)
ISBN 978-1-4665-6123-6

MIMO Processing for 4G and Beyond: Fundamentals and Evolution
Mário Marques da Silva and Francisco A. Monteiro (Editors)
ISBN 978-1-4665-9807-2

Network Innovation through OpenFlow and SDN: Principles and Design
Fei Hu (Editor)
ISBN 978-1-4665-7209-6

Opportunistic Mobile Social Networks
Jie Wu and Yunsheng Wang (Editors)
ISBN 978-1-4665-9494-4

Physical Layer Security in Wireless Communications
Xiangyun Zhou, Lingyang Song, and Yan Zhang (Editors)
ISBN 978-1-4665-6700-9

SC-FDMA for Mobile Communications
Fathi E. Abd El-Samie, Faisal S. Al-kamali, Azzam Y. Al-nahari, and Moawad I. Dessouky
ISBN 978-1-4665-1071-5

Security for Multihop Wireless Networks
Shafiullah Khan and Jaime Lloret Mauri (Editors)
ISBN 978-1-4665-7803-6

Self-Healing Systems and Wireless Networks Management
Junaid Ahsenali Chaudhry
ISBN 978-1-4665-5648-5

The State of the Art in Intrusion Prevention and Detection
Al-Sakib Khan Pathan (Editor)
ISBN 978-1-4822-0351-6

Wi-Fi Enabled Healthcare
Ali Youssef, Douglas McDonald II, Jon Linton, Bob Zemke, and Aaron Earle
ISBN 978-1-4665-6040-6

Wireless Ad Hoc and Sensor Networks: Management, Performance, and Applications
Jing (Selina) He, Shouling Ji, Yingshu Li, and Yi Pan
ISBN 978-1-4665-5694-2

Wireless Sensor Networks: From Theory to Applications
Ibrahiem M. M. El Emary and S. Ramakrishnan (Editors)
ISBN 978-1-4665-1810-0

ZigBee® Network Protocols and Applications
Chonggang Wang, Tao Jiang, and Qian Zhang (Editors)
ISBN 978-1-4398-1601-1

AUERBACH PUBLICATIONS
www.auerbach-publications.com
To Order Call: 1-800-272-7737 • Fax: 1-800-374-3401
E-mail: orders@crcpress.com

Power Efficiency in Broadband Wireless Communications

Pooria Varahram
Somayeh Mohammady
Borhanuddin Mohd Ali
Nasri b. Sulaiman

CRC Press
Taylor & Francis Group
Boca Raton   London   New York

CRC Press is an imprint of the
Taylor & Francis Group, an **informa** business

CRC Press
Taylor & Francis Group
6000 Broken Sound Parkway NW, Suite 300
Boca Raton, FL 33487-2742

© 2015 by Taylor & Francis Group, LLC
CRC Press is an imprint of Taylor & Francis Group, an Informa business

No claim to original U.S. Government works

Printed on acid-free paper
Version Date: 20140613

International Standard Book Number-13: 978-1-4665-9548-4 (Hardback)

Library of Congress Cataloging-in-Publication Data

Varahram, Pooria.
    Power efficiency in broadband wireless communications / authors, Pooria Varahram, Somayeh Mohammady, Borhanuddin Mohd Ali, Nasri B. Sulaiman.
        pages cm
    Includes bibliographical references and index.
    ISBN 978-1-4665-9548-4 (hardcover : alk. paper) 1. Wireless communication systems--Energy conservation. 2. Wireless communication systems--Energy consumption. 3. Engineering economy. I. Title.

TK5103.2.V38 2015
621.384--dc23   2014022605

Visit the Taylor & Francis Web site at
http://www.taylorandfrancis.com

and the CRC Press Web site at
http://www.crcpress.com

Contents

Preface

This book focuses on the study and development of one of the most advanced topics in broadband wireless communications systems: power efficiency and power consumption in wireless communications systems, especially of mobile devices. Hence, the main focus of this book is on the most recent techniques for the conservation of power and increase in power efficiency. This is an important topic that has been addressed in recent consumer electronics publications where the main challenge is to prolong the battery life of mobile devices and reduce the system costs. To achieve both of the aforementioned objectives in consumer devices, scrutinizing recent technologies and their impact on the consumer devices is vital. We will discuss each of these topics, first by introducing the main physical layer components in recent wireless communications systems and their shortcomings, and then proposing appropriate solutions to overcome these shortcomings.

Because the most costly device in a communication system that has a direct impact on power efficiency and power consumption is the power amplifier, we will study the behavior and characteristics of different classes of power amplifiers in detail. The subsequent topics will deal directly or indirectly to the power amplifiers. The nonlinear behavior of the power amplifier has made it an interesting research area for the last 50 years. Although the nonlinear characteristics of the power amplifier are known, developing its analytical model was a major challenge in the past. There are different power amplifier models that have been researched, including the Saleh model, Taylor series, Ghorbani model, Rapp model, Volterra series, and memory polynomial. The latter is the most optimum model that has been derived; it can be applied for a wide variety of power amplifiers in wireless communications systems. The main feature of this model is that it considers electrical or short-term memory effects. In this book, we will study and analyze this model as well.

This book is organized as follows: Chapter 1 introduces the mobile cellular communications system and reviews its evolution. It also discusses multiplexing techniques and the most optimum multiplexing solution, which includes spectral efficiency and interference mitigation.

Chapter 2 examines orthogonal frequency division multiplexing (OFDM) signal generation and formulation. As the most recent multiplexing technique, OFDM and its advantages and disadvantages are discussed.

Chapter 3 explains the power amplifier characteristics in wireless communications systems. It is the most costly device in communications systems and its behavior needs to be understood. The main parameters that define a power amplifier, AM-AM and AM-PM, are introduced here and the challenge of nonlinearity and its impact on the output spectrum as well as in-band distortion are shown. Several classes of power amplifiers and the most reliable class in terms of linearity and efficiency are introduced.

In Chapter 4, one of the main drawbacks of OFDM systems, peak-to-average power ratio (PAPR), is introduced and several PAPR techniques are discussed. Moreover, simulation results of a new PAPR technique are included. Two of the most promising techniques, selected mapping and partial transmit sequences, are also presented and explained.

In Chapter 5, the implementation of an optimum PAPR technique and its respective hardware platform is introduced. The main measurement parameter and comparison between simulations will be carried out.

Chapter 6 continues with an introduction to power amplifier linearization to increase power efficiency and hence save system costs and power dissipation. Different linearization techniques are introduced and a scheme to overcome the drawbacks of conventional techniques is also presented.

Chapter 7 describes the implementation scenario of digital predistortion, the most promising linearization technique.

Chapter 8 presents some experimental demonstrations of digital predistortion when the device under test (DUT) is in the loop.

MATLAB® is a registered trademark of The MathWorks, Inc. For product information, please contact:

The MathWorks, Inc.
3 Apple Hill Drive
Natick, MA 01760-2098 USA
Tel: 508-647-7000
Fax: 508-647-7001
E-mail: info@mathworks.com
Web: www.mathworks.com

1

EVOLUTION OF MULTIPLEXING TECHNIQUES IN WIRELESS COMMUNICATIONS SYSTEMS

1.1 Introduction

The main purpose of a communication system is to transmit data or information from one point to another or from a source to a destination. A general communication system is comprised of three parts: a transmitter, channel, and receiver. The channel is the medium where the data propagates and it can be wired or wireless. In this book we will only cover the wireless channel, hence in the transmitter there is an antenna to send the data to a specific distance. On the receiver side, there is an antenna to receive a signal, which normally is low in strength and needs to be further amplified. The main blocks in the transmitter of a general communication system include the coding, interleaving, modulation, digital-to-analog converter (DAC), mixer, filter, power amplifier, and antenna. In the receiver, however, there is a low noise amplifier (LNA), a band-pass filter (BPF), downconversion, demodulation, analog-to-digital conversion (ADC), decoder, and interleaver. It should be noted that for different generations of communication systems and different applications of broadband communication systems, the transmitter and receiver blocks are different than those mentioned earlier. The main concentration in this book is on the most costly device used in communication systems, the power amplifier (PA). Sometimes it is denoted with HPA, which stands for high power amplifier. This is because in current wireless communications systems the need for coverage and transmission to long distances makes it necessary to design PAs with high power.

As the power amplifier is a costly device, optimum usage of its power is desired. Initially though, the behavior of this device and its working conditions should be discussed. There are different working classes of PAs, such as class AB, A, B, C, D, E, and F. Here, we will study these classes and explain which one is most suitable in wireless applications. There is always a trade-off to select between the power amplifier classes. Some classes have high efficiency and some have high linearity. Because it is necessary to have a power amplifier with high linearity and high efficiency in wireless communications systems, the most suitable one is class AB, which is a good compromise between linearity and efficiency. There is another type of PA that was recently introduced, this is called Doherty, and is comprised of classes AB and C. It has a high linearity and efficiency among other power amplifiers (Chari, 2007).

Wireless communications systems have experienced fast growth in the past few years. A basic communication system includes radio frequency front-end and baseband processing. From an application point of view, it can be categorized as unicasting or broadcasting. Unicasting is used for symmetric applications, such as telephony and wireless access to data. However, broadcasting is used to distribute audio and video over reliable channels. An important application of broadband communications is cellular networks, which nowadays is in its fourth generation.

1.2 Evolution of Mobile Cellular Networks

The cellular system is one of the most successful applications of wireless communication. In the cellular concept, each cell consists of a base station (BS) and supports several mobile stations within its coverage area. These cells are hexagonal and have a specific frequency band that is arranged to be different from its neighboring cells. The evolution of mobile cellular systems started from the first generation, which used analog technology, and ever since has progressed over several generations to what is now referred to as fourth-generation cellular systems (4G). Researchers, manufacturers, and regulators are already preparing for 5G. Here, we briefly describe this evolution (Goldsmith, 2005).

1.2.1 First-Generation Cellular Systems

The first generation of cellular systems was designed in the early 1960s, but was not launched until the 1980s due to regulatory delays. It was based on the frequency division multiplexing access (FDMA) technique to transmit analog data for long distances. The first generation could only carry voice services over circuit switching.

1.2.2 Second-Generation Cellular Systems

Following the development of communication and electronic technology and the introduction of digital systems, the second generation of cellular systems was developed in the 1990s. In fact, a key difference between 1G and 2G cellular systems is digital communication. Digital communication comprises of a digital-to-analog converter (DAC) and an analog-to-digital converter (ADC), both of which are located in the transmitter and receiver of the communication systems, respectively. Some of the main advantages of 2G systems over 1G are listed as follows:

1. Better security
2. Better communication quality
3. Higher spectrum efficiency

Furthermore, 2G systems can achieve longer battery life in addition to a higher data rate, higher bandwidth, and cheaper equipment compared to analog systems.

Both 2G and 1G use circuit switching technology. 2G systems include: Global System for Mobile (GSM) communication, which was used in Europe; Personal Digital Cellular (PDC); and IS-95.

In general, 2G systems use time division multiplexing access (TDMA), which is more bandwidth efficient than FDMA.

1.2.3 Third-Generation Cellular Systems

Due to the tremendous growth in the number of mobile subscribers and the demand for higher bandwidth and high data rates, cellular mobile systems have been further developed to satisfy these demands. The third generation (3G) of cellular systems was introduced at the beginning of the 21st century. The main feature of 3G systems is its packet switching for data transmission. The 3G standards were defined as International Mobile Telecommunication (IMT-2000) by the International Telecommunication Union (ITU). IMT-2000 aims to support high speed voice and data services, seamless roaming, and delivery of services.

3G systems use code division multiple access (CDMA), which can use bandwidth and data rates effectively. There are three types of CDMA in 3G. They are wideband CDMA (WCDMA), CDMA2000, and time division synchronous CDMA (TD-SCDMA), which were proposed by three different trade blocs of Europe, the United States, and China. 3G networks can support data rates from 384 kbps to 2 Mbps, while 2G networks support data rates ranging from 9.6 kbps to 28.8 kbps.

1.2.4 Future Broadband Wireless Communication

In the future, different wireless communication standards are expected to be integrated into one unique communication system. This new system is in fact categorized as a fourth-generation (4G) system. It includes comprehensive and secure IP-based solutions (Motorola, 2007). However, a new generation of cellular wireless communications systems is foreseen and is going to be rolled out, which is the fifth generation or 5G. This new system is based on generalized frequency division multiplexing (GFDM). In the next section we will talk more about this new multiplexing technique.

Some of the main features of 4G technology include:

1. Efficient modulation technique—A new modulation technique called orthogonal frequency division multiplexing (OFDM) is used in the 4G systems.
2. Advanced antenna technique—A multiple antenna technique called multiple input multiple output (MIMO) is employed to combat various types of channel fading.
3. The use of advanced error coding and decoding techniques such as turbo coding and low density parity check (LDPC) codes.
4. Use of flexible wireless access technologies such as orthogonal frequency division multiple access (OFDMA) and multiple-carrier code division multiple access (MC-CDMA).

1.3 Evolution of Multiplexing Techniques

1.3.1 Frequency Division Multiplexing Access (FDMA) Technique

FDMA is a technology that assigns specific channels to different users by means of the multiple access technique. FDMA is used in several applications, including satellite communications, cable, and terrestrial radio. Hence, in FDMA the total bandwidth is divided into N narrowband channels called subchannels, as shown in Figure 1.1a. In order to do that a band-pass filter with specific stopband attenuation is required. Moreover, to minimize the adjacent channel interference (ACI) a guard band is allocated between two adjacent spectra. This is due to the deviation caused by local oscillators. The main advantages of FDMA is its low transmitting power and less complex channel equalization. However, its main drawback is that in a cellular system the implementation of a couple of modulators and demodulators at the base station leads to a higher deployment cost.

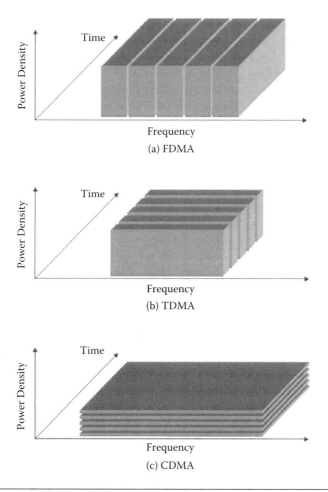

Figure 1.1 Comparison of multiplexing techniques.

1.3.2 Time Division Multiplexing Access (TDMA) Technique

TDMA is a more popular multiple access technique that is used in several communication systems. In TDMA, the spectrum is divided into time slots and each time slot is allocated to one user as shown in Figure 1.1b.

In TDMA, data transmitted in the buffer and burst format result in noncontinuous transmission. Despite FDMA being based on analog FM, in TDMA digital data and digital modulation must be used. Some features of TDMA are as follows:

1. In TDMA a single carrier frequency is shared with several users.
2. Due to the discontinuous transmission in TDMA, a hand-off process is much simpler.
3. An adaptive equalizer is usually necessary in TDMA systems, which increases the system cost compared to FDMA.
4. High synchronization overhead is required in TDMA systems due to burst transmission.

1.3.3 Code Division Multiple Access (CDMA) Technique

In CDMA systems, the spread signal technique is used, which shares the same frequency band and time, as shown in Figure 1.1c. Each spread signal can be made unique by means of unique patterns called code words, and they are designed to be orthogonal such that they do not interfere with one another. Some of the main characteristics of CDMA systems include:

1. Multipath selective fading may be substantially reduced due to the spreading of the signal over a large spectrum
2. Very high channel data rate

However, self-jamming is one of the main drawbacks of CDMA.

1.3.4 Orthogonal Frequency Division Multiplexing (OFDM) in 4G

OFDM is a multicarrier modulation technique that can deliver high data rates in a fading channel. The idea is to divide an input high bit rate data stream into several parallel low bit rate signals, following modulation and inverse fast Fourier transform (IFFT), which is the core of OFDM systems. The main advantage of OFDM over FDMA is the orthogonality of adjacent subcarriers, which makes OFDM more bandwidth-efficient than a plain FDMA. At the receiver only, a trivial frequency domain equalizer and a fast Fourier transform to convert a time domain signal to the frequency domain is required. The size of the FFT needs to be carefully set; a larger FFT size provides better immunity against frequency selective fading. However, OFDM is more sensitive to Doppler shifts and in addition the complexity and system costs will be increased if the number of IFFT is large. In order to combat intersymbol interference (ISI) in OFDM systems, a technique called cyclic prefix is applied where a tail of a previous OFDM symbol is copied to its head and the amount of the data that has to be copied depends on the delay spread of the channel.

1.3.4.1 OFDM Pros and Cons   As with most systems, OFDM has several advantages and disadvantages over other solutions for high-speed transmission. One of the advantages is low computational complexity. Previous wireless technologies, such as 3G and 2G, use TDMA and CDMA multiplexing techniques where due to the high probability of ISI at the receiver, the bit error rate (BER) was a major concern. Hence, a robust equalization technique has to be embedded in those systems to successfully recover the transmitted signal. The literature shows that the complexity of implementing an equalizer is very high and therefore significantly raises the cost of the system.

Despite the advantages, OFDM has some main drawbacks. High peak-to-average power ratio (PAPR) is a major problem with OFDM signals. This is due to the nature of OFDM, which causes time domain superposition of the input signal. At times, the individual signals add up including high peaks and at times the individual signals subtract each other. This makes the peak of the overall signals high compared to the expected value or average. If this high PAPR signal is passed through a power amplifier, it forces the power amplifier to work at its nonlinear region, which is not a desirable operating region. As a result, there are several PAPR techniques that help to overcome this main drawback.

1.4 Key Technologies

To design a communication system that provides optimum bandwidth allocation, with a high data rate and reliable, secure, and high quality of service, there are some technologies that have been recently introduced, which are discussed next.

1.4.1 Generalized Frequency Division Multiplexing (GFDM)

GFDM is a multicarrier system that digitally implements the classical filter band approach. Cyclic prefix (CP) insertion is used to allow for low complex equalization at the receiver side. For this purpose, filters with the raised cosine (RC) and root-raised cosine (RRC) shape have proven suitable properties. Some of the features of the GFDM compared to OFDM is its lower PAPR due to the fact that GFDM makes the subcarriers into one subcarrier by using a discrete Fourier transform (DFT) block, hence the cumulative PAPR will significantly reduce. In addition, the synchronization problem that was critical in OFDM systems is resolved in GFDM systems. However, the complexity is increased compared to OFDM systems. Figure 1.2 shows the block diagram of the GFDM transmitter system in detail.

The receiver, as shown in Figure 1.3, is an inverse of the transmitter where there is a low noise amplifier (LNA) at the front end and downconversion and filtering and FFT and IFFT to recover original data.

1.4.2 Multiple Input Multiple Output (MIMO)

MIMO multiple antenna technology has been applied in new generations of wireless systems. This technology has already entered into the 3G and LTE systems. It is also employed in the IEEE 802.11n WLAN standard.

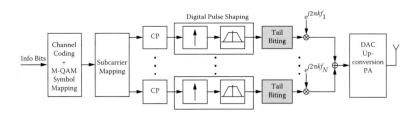

Figure 1.2   Principle of the GFDM digital transmitter.

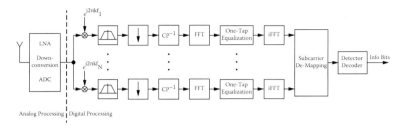

Figure 1.3 Principle of the GFDM digital receiver.

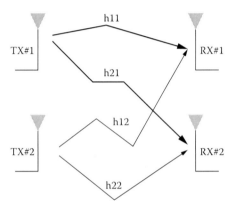

Figure 1.4 Basic 2 × 2 MIMO channel configuration.

The main advantage of MIMO is that the signal at the transmit and receiver antennas is combined in specific ways, which improve the bit error rate or increase the data rate. This performance is improved by exploiting the array gain, interference reduction, spatial multiplexing, and diversity gain.

The basics of MIMO operation can be understood by using a static four-port network to represent the channel, as shown in Figure 1.4. In this figure, signals from two different antennas are transmitted and at the receiver two antennas received the signal.

In the ideal scenario, and in order to use the same frequency and time, an isolated connection from transmitter 1 to receiver 2 and transmitter 2 to receiver 1 would need to be established. However, in practice this is not possible as there would be some coupling between the signals once they were transmitted.

A key point about MIMO is there must be at least as many receiving antennas as there are transmitted data streams (Kalis et al., 2008).

1.4.3 Space Time and Space Frequency Transmission over MIMO Networks

In this section, the physical layer is delineated for integrating MIMO technology in WCDMA or OFDMA networks. In this regard, some consideration should be given to the complexity of the transmitter and receiver signaling burden, the channel measurements at the receiver, and data rates. Several researchers have addressed the deployment of MIMO in 3G networks. A very popular transmission technique is per antenna rate control (PARC). In this technique, a high data rate stream is demultiplexed into a number of T substreams, where T is the number of transmit antennas. Following that the data is encoded and modulated from different options, QPSK (quadrature phase-shift keying) and 16 quadrature amplitude modulation (QAM), and then demultiplexed. At the receiver side, the signal-to-interference-plus-noise ratio (SINR) is measured separately (Li et al., 2008).

1.5 Summary

In this chapter, a review of multiplexing techniques in wireless communications systems and the evolution of mobile telecommunications generations were introduced. The evolution of multiplexing techniques leads to a robust technique called orthogonal frequency division multiplexing (OFDM), which overcomes the problem of intersymbol interference (ISI) of fading channels without requiring complex equalizers. Finally, in this chapter a different type of system, MIMO, was introduced.

References

Chari, M. R., and F.-Y. Ling. 2007. FLO physical layer: An overview. *IEEE Transaction on Broadcasting* 53:145–160.

Goldsmith, A. 2005. *Wireless Communications.* Cambridge University Press.

Kalis, A., Kanatas, A., and Papadias, C. 2008. A novel approach to MIMO transmission using single RF front end. *IEEE Journal on Selected Areas in Communication* 26:972–978.

Li, H., Liu, B., and Liu, H. 2008. Transmission schemes for multicarrier broadcast and unicast hybrid systems. *IEEE Transaction on Wireless Communications* 7:4321–4330.

Motorola Inc. 2007. Long-term evolution (LTE): A technical overview. Motorola Technical White Paper.

2

Orthogonal Frequency Division Multiplexing Theory

2.1 Introduction

It is clear from the previous chapter that orthogonal frequency division multiplexing (OFDM) achieves high spectral efficiency, high data rates, and dynamic allocation of bandwidth to users without complex implementation. As a result, various current and future wireless systems such as IEEE 802.11g, s, n (WiFi); Long-Term Evolution (LTE); Worldwide Interoperability for Microwave Access (WiMAX); and Digital Video Broadcast (DVB) have adopted OFDM as their modulation technique. Among all these technologies, WiMAX is the most popular because of its high bandwidth (i.e., >10 MHz), coverage, cost effectiveness, and ability to match to any network type. One important part of the physical layer in WiMAX is the OFDM; it can be considered as a special form of multicarrier modulation (MCM). The OFDM process is similar to the frequency division multiplexing (FDM) signal, although there are some differences. FDM runs on a single carrier signal in which signals are divided into frequency bands with some guard intervals to prevent interference. However, in OFDM the subcarriers are orthogonal to each other, thus interference is not an issue and the bandwidth is used more efficiently.

The OFDM system employs inverse fast Fourier transform (IFFT) and fast Fourier transform (FFT) at the transmitter and receiver, respectively, for modulation and demodulation. The IFFT processor provides orthogonality between adjacent subchannels. However, despite all the advantages, OFDM is beset with some design challenges that limits its performance. One of the challenges is the high peak-to-average power ratio (PAPR), which is described in Chapter 4. In this chapter, we analyze OFDM modulation and investigate its characteristics.

2.2 History of OFDM

In wireless communications, the signal transmitted from the source usually faces many obstructions such as buildings, mountains, trees, cars, and other obstacles or objects before it reaches the destination. Therefore, it experiences reflection, attenuation, scattering, and refraction, and the combined effect at the receiver causes the signal energy to fluctuate following a certain distribution depending on the environment. This is generally known as a multipath effect or multipath channel response. If the data is transmitted at high symbol rates, the bandwidth of the signal becomes wider and its channel response can be assumed to be fixed (flat). This wide bandwidth is known as the coherence bandwidth of the channel (Brooks et al., 2001).

These fading effects degrade the signal and usually a portion of symbol energy is also lost. This effect is known as intersymbol interference (ISI). To eliminate ISI, an adaptive digital filter, known as an equalizer, is usually employed in the receiver. When the equalizer is employed, the combined response of the channel and equalizer can be assumed constant within the signal bandwidth. However, the equalizer suffers from a complicated design and many other limitations. One limitation is that the estimation of the channel response has a significant effect on the performance of an equalizer. For example, if the channel response is longer than the equalizer's length, the equalizers will perform poorly.

To avoid using complicated equalizers, it was proposed in the 1960s that the main data be split into parallel sections known as streams. Therefore, each of these streams has a reduced symbol rate and bandwidth. When the main bandwidth is divided among a large number of streams, the bandwidth of each stream will be less than the channel coherence bandwidth. As a result, the channel response for each stream will be flat. Therefore, less compensation process is required at the receiver side.

The WiMAX system is one of the most popular applications of this high-efficiency signal since it can satisfy a variety of user demands, such as high bandwidth (i.e., >10 MHz) and coverage as well as being a cost-effective system that can be matched to any network. The first form of WiMAX technology was designed in 2003 following the standard of IEEE 802.16a in which the frequency range of operation was between 2 and 11 GHz. Further revisions were made in 2004 and 2005, which enabled high speed mobility applications. Eventually, two main categories of WiMAX signals were defined: one for point to point(s) application known as a fixed application that uses IEEE 802.16d or 802.16-2004 standard; and the most prevalent, which is called mobile WiMAX that uses IEEE 802.16e or 802.16-2005.

The final release of the WiMAX standard is the 802.16m/n in 2011, which supports high-speed mobile users and provides more security features. This was drafted to meet the International Mobile Telecommunication (IMT) advanced specifications of the International Telecommunications Union Standardization Sector (ITU-T) or 4G.

2.3 OFDM Blocks

A simplified OFDM transmitter is shown in Figure 2.1, where the input data stream with a high data rate of R_s is introduced to a serial-to-parallel converter. The serial-to-parallel converter divides the high rate stream into K separate substreams, in which each substream has a lower rate of Rs/K symbols per second.

An important block of an OFDM transmitter is the inverse discrete Fourier transform (IDFT) function, which creates the orthogonality feature between subcarriers and prevents interference

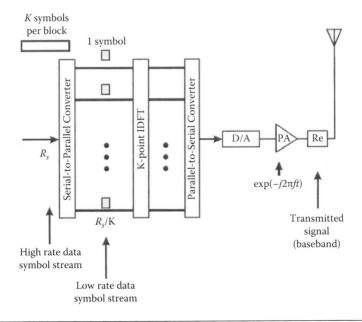

Figure 2.1   Block diagram of an OFDM transmitter.

among the carriers. This leads to further savings of bandwidth and enhances bandwidth efficiency. The size of the IDFT function, K, depends on the standard used in the transmitter. For example, in the WiMAX standard, it might be 64, 128, 256, 512, 1024, or 2048 depending on the number of subcarriers.

After the IDFT block, the K symbols have to be transformed back to the series format. Then the signal is given to a digital-to-analog converter (DAC) or D/A, which creates an analog signal, and then it is introduced to the amplification stage (the power amplifier [PA]) and antenna.

Then the signal passes through the channel and is recovered at the receiver. In ideal conditions for the wireless channel, the signal path is the line of sight (LOS) with no obstacles; however, in actuality the signal will experience multipaths that might be diffracted, refracted, reflected, or absorbed by the atmosphere and obstructions, and therefore signal protection becomes crucial and costly.

The main advantage of the OFDM is relative robustness against fading caused by a multipath environment compared to single carrier signals. A multipath channel has an added drawback that is called intersymbol interference (ISI). In ISI, a symbol interferes with subsequent symbols and reduces the communication reliability. Therefore, in the design of the transmitter and receiver, minimizing the effects of ISI should be considered. In the OFDM system, the ISI is prevented by using a cyclic prefix (CP), which copies the last portion of the data symbol to the beginning of the symbol, as shown in Figure 2.2. The total time length of the OFDM signal symbol is $T_{sym} = T_b + T_g$, where T_b is the useful time and T_g is the time of CP. According to the literature (for example, Roca, 2007), the ratio between Tg and Tb would be at least 1/32.

Since the CP is useless at the receiver, it can be discarded before the FFT process in time domain, as shown in Figure 2.3. It should be noted that the addition of the CP does not have any effect on PAPR (Baxley and Zhou, 2007).

As a result of channel noise and multipath fading, the signal received is not the same as the transmitted signal and there are usually various corruptions that are changeable with time, place, and weather; therefore, channel modeling and channel estimation are critical for detecting the original data. As the OFDM receiver block diagram in Figure 2.3 shows, every operation that has been completed in the transmitter is reversed on the receiver side.

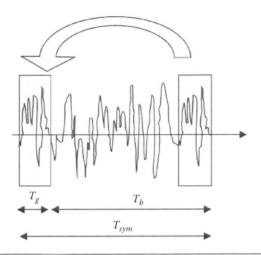

Figure 2.2   The cyclic prefix of the OFDM signal.

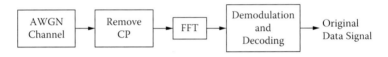

Figure 2.3   The block diagram of the OFDM receiver. (From Roca, A. 2007. Implementation of a WiMAX simulator in Simulink. Graduate engineer dissertation, Technical University of Vienna, Spain. With permission.)

The signal is then transformed to the frequency domain by using an FFT processor and after a demodulation block, the original data can be retrieved.

2.4 OFDM Mathematical Analysis and Measurements

If we assume a random serial data input of $D = \{0, 0, 0, 1, 1, 0, 1, 1, \ldots\}$ as an example to illustrate the construction of the OFDM signal, then the paralleled signal can be presented as

$$D_1 = \{0, 0\}$$

$$D_2 = \{0, 1\}$$

$$D_3 = \{1, 0\} \tag{2.1}$$

$$D_4 = \{1, 1\}$$

$$\ldots$$

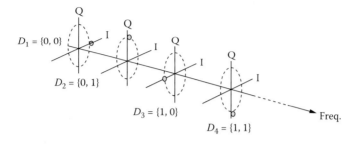

Figure 2.4   Constellation diagram of an OFDM symbol.

Then, D_i, $(i = 1, 2, \ldots, N)$ in Equation (2.1) is given to the modulation block where in-phase and in-quadrature (I and Q) constellations are generated. Here, quadrature phase-shift keying (QPSK) is used to map two bit symbols into the constellations, X_i, as shown in Figure 2.4.

$$X_1 = 1 + 0j$$
$$X_2 = 0 + 1j$$
$$X_3 = (-1) + 0j \qquad (2.2)$$
$$X_4 = 0 + (-1)j$$

$$\ldots$$

After the modulation, the signal is introduced to the IFFT block. The IFFT block takes frequency domain data and generates corresponding time domain samples (Brooks et al., 2001; Roca, 2007). These samples and their adjacent samples are orthogonal to each other, which means that their dot product will be zero. Following the IFFT process, the OFDM signal can be shaped by converting time domain samples sequentially by the parallel-to-serial converter. This signal, X, can be presented as Equation (2.3):

$$X = \{0.25 + j0.75, -0.25 - j0.25, 0.25 + j0.25, -0.25 - j0.25, \ldots\} \quad (2.3)$$

According to the IEEE standard (IEEE STD 802.16e–2005), the time domain view of the OFDM signal is as shown in Figure 2.3. This example shows a typical OFDM signal with an IFFT length of 256. According to the IEEE standard, 55 subcarriers are used as a guard to prevent interferences and for the purpose of keeping the spectrum balanced.

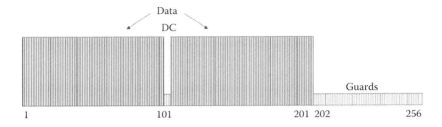

Figure 2.5 The structure of an OFDM signal. (From Roca, A. 2007. Implementation of a WiMAX simulator in Simulink. Graduate engineer dissertation, Technical University of Vienna, Spain. With permission.)

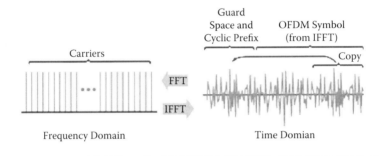

Figure 2.6 The frequency and time domain format of an OFDM signal.

As shown in Figure 2.5, the DC zero subcarrier is placed in an index of 101. Guard zeros are at the end with an index of 202 to 256. These guards decrease emissions in adjacent frequency channels, however, the guard will be halved for both sides of the signal. Another presentation of the frequency domain and time domain OFDM signal is shown in Figure 2.6.

It can be seen that the OFDM signal in time domain presents some very high peaks. These sharp maximums are due to the constructive characteristics of sinusoid waves. Since there is also a destructive phenomena between sinusoid waves, the average OFDM signal might be very low or zero. Therefore, the ratio between the maximum and average OFDM signal would be very high. This ratio is called the peak-to-average power ratio (PAPR) or crest factor (CF).

As mentioned earlier, high PAPR is the major practical problem involving OFDM. If $A = (A_0, A_1, ..., A_{(N-1)})$ is a modulated data sequence of length N in the time interval $(0,T)$, where A_i is a symbol from a signal constellation and T is the OFDM symbol duration, then the N carriers of the OFDM envelope are given by (Ochiai, 2003)

$$s(t) = \sum_{n=0}^{N-1} A_n e^{j\omega_0 nt} \tag{2.4}$$

where, $\omega_0 = 2\pi/T$ and $j = \sqrt{-1}$. The PAPR of the transmitted signal $s(t)$, defined in Equation (2.4), is the ratio of the maximum power and the average power of the signal and can be defined by

$$PAPR(A) = \frac{S_{max}^2}{E\left[|S_n|^2\right]} = \frac{S_{max}^2}{N} \tag{2.5}$$

where E denotes the expectation operator that calculates the average of the signal. The continuous time PAPR of $s(t)$ is approximated using the discrete time PAPR from OFDM signal samples (Rajbanshi and Veeragandham, 2004).

Another parameter is the crest factor, which is widely used in the literature, and defined as the square root of the PAPR:

$$Crest\,Factor(C.F.) = \sqrt{PAPR} \tag{2.6}$$

Basically, the function that is used to study the effects of crest factor reduction (CFR) methods is called the complementary cumulative distribution function (CCDF). It denotes the probability that the PAPR of a data symbol exceeds a predefined threshold as expressed by (Han and Lee, 2005; Heo et al., 2009):

$$probability\,(PAPR > z) = Prob.(S_{max}^2 > zN)$$

$$= 1 - F_{S_{max}^2}(zN) = 1 - (1 - \exp(-z))^N \tag{2.7}$$

$$F(z) = 1 - \exp(z)$$

where N is the number of subcarriers and z is the threshold. Essentially, this probability function is used as a graph to determine the ability of an algorithm in reducing the PAPR of the OFDM signal, and the PAPR is usually compared to the unmodified OFDM signal at 0.01%

CCDF, which is shown by the 10^{-4} CCDF in the horizontal vector of graphs. A typical OFDM signal without any PAPR reduction technique has about 8 dB to 13 dB PAPR at 10^{-4} CCDF (Raab et al., 2002). Therefore, when a PAPR reduction technique has to be applied to the OFDM system, it is expected to reduce the 13 dB PAPR to some lower value. According to the IEEE standard (IEEE STD 802.16e–2005), the reduction should be at least 3 dB.

It should be noted that any modification in the transmitted signal will affect the accuracy of receiving a signal, therefore, the error occurring on the number of correctly received signals, known as bit error rate (BER), is a critical analysis in this field.

The additive white Gaussian noise (AWGN) channel is used to estimate channel effects and measure the BER to assess transmission performance (Kumar et al., 2007; Stewart et al., 2009). The BER shows the percentage of bits that is received with errors relative to the total number of bits received in a transmission. This can be computed from the bit error probability given by (Le Goff et al., 2008; Sen et al., 2008):

$$P_{be} >> \frac{4(\sqrt{V}-1)}{k\sqrt{V}} Q\left(\sqrt{\frac{3k}{(V-1)} \cdot \frac{E_b}{N_o}} \right) \tag{2.8}$$

where V is the modulation order, for QPSK modulation $V = 4$, while $Q(.)$ is the Gaussian Q function defined as

$$Q(y) = erfc\left(\frac{y}{\sqrt{2}} \right) \tag{2.9}$$

where $erfc(.)$ is the matching error function, y is the amplitude, and $k = log_2(V)$ refers to the number of bits per symbol.

2.5 Summary

In this chapter, we have described the mechanism of OFDM and the problem of PAPR. In summary, the OFDM transmission scheme has the following key advantages:

- High bandwidth efficiency referred to as spectrum efficiency or spectral efficiency. This shows the information rate that can be transmitted over a certain bandwidth. As a result of the orthogonal feature and the overlapping spectra of the subcarriers, OFDM can achieve higher bandwidth efficiency than a single carrier system such as a simple FDM system (Paulraj et al., 2004; Visser and Bar-Ness, 1999).

- Cost effective and simple implementation by using the IFFT processor. The OFDM system can be implemented using field programmable gate arrays (FPGAs) and digital signal processors (DSPs). The IFFT provides the orthogonality characteristic of the produced carrier signals. Since all signal processing is made digitally in the frequency domain, the cost of implementing one IFFT is reasonably low (Kadiran, 2005).
- Enhanced immunity against frequency selective fading than single carrier systems. By dividing the main channel into the number of narrowband subcarriers, in the OFDM system it is easier to compensate the effects of frequency selective fading. The effect of fades can be assumed to be steady in each of the narrowband channels (Rajbanshi and Veeragandham, 2004).
- Low hardware complexity receiver. The implementation of the OFDM modulators and demodulators is practical for inexpensive transmitters and receivers. Since all the carriers are orthogonal to each other, the only necessary process at the receiver is to perform an FFT and demodulation (Cui et al., 2006). This process is more efficient and less complex than the conventional complicated equalizers used to retrieve the signal instead of FFT.
- Highly adaptive and flexible in terms of link and scalability. The data rate, modulation order, and the IFFT size can be optimized in order to satisfy application requirements. The OFDM is already applied in Worldwide Interoperability for Microwave Access (WiMAX), Long-Term Evolution (LTE) uplink, 4G, European Digital Audio/Video Broadcasting (DAB/DVB), Asymmetric Digital Subscriber Line–Discrete Multi-Tone (ADSL-DMT), and so on (Zhang and Thibault, 2010).

In addition to the advantages of OFDM systems, OFDM has some drawbacks:

- High PAPR. As highlighted earlier, an OFDM signal is generated as a summation of many sinusoids. At some time instances, due to constructive characteristics, this sum is large and at other times it is small, which means that the peak value of the signal will be substantially larger than the average value. This will force the power amplifier to work harder to avoid distortion, which reduces the efficiency of the transmitter.

- High sensitivity to carrier frequency offset. This is also known as an error of carrier frequency synchronization (Visser and Bar-Ness, 1999). It might cause loss of the orthogonality between carriers, and hence intercarrier interference (ICI) cannot be fully prevented.

The first drawback, which is high PAPR, will be discussed in the next chapter.

References

Baxley, R. J., and Zhou, G. T. 2007. Comparing selected mapping and partial transmit sequence for PAR reduction. *IEEE Transactions on Broadcasting* 53(4):797–803.

Brooks, A. C., Hoelzer, S. J., Stewart, T. L., and Ahn, I. S. 2001. Design and simulation of orthogonal frequency division multiplexing (OFDM) signaling. Bradley University, Peoria, IL.

Cui, X., Yu, D., Sheng, Sh., and Cui, X. 2006. A CORDIC demodulator platform for digital-IF receiver. Proceedings of 8th International Conference on Solid-State and Integrated Circuit Technology (ICSICT '06), Beijing, October 23–26.

Han, S. H., and Lee, J. H. 2005. An overview of peak-to-average power ratio reduction techniques for multicarrier transmission. *IEEE Wireless Communications* 12(2):56–65.

Heo, S., Joo, H., No, J., Lim, D., and Shin, D. 2009. Analysis of PAPR reduction performance of SLM schemes with correlated phase vectors. Proceedings of the IEEE International Symposium on Information Theory (ISIT 2009), Seoul, June 28–July 3.

Institute of Electrical and Electronics Engineers (IEEE). IEEE STD 802.16e™ 2005, IEEE Standard for Local and Metropolitan Area Networks.

Kadiran, K. A. B. 2005. Design and implementation of OFDM transmitter and receiver on FPGA hardware. Unpublished doctoral dissertation, Universiti Teknologi Malaysia (UTM).

Kumar, N. S. L. P., Banerjee, A., and Sircar, P. 2007. Modified exponential companding for PAPR reduction of OFDM signals. Proceedings of IEEE Wireless Communications and Networking Conference (WCNC 2007), Hong Kong, March 11–15.

Le Goff, S. Y., Khoo, B. K., Tsimenidis, C. C., and Sharif, B. S. 2008. A novel selected mapping technique for PAPR reduction in OFDM systems. *IEEE Transactions on Communication* 56(11):1775–1779.

Ochiai, H. 2003. Performance analysis of peak power and band-limited OFDM system with linear scaling. *IEEE Transactions on Wireless Communications* 2(5):1055–1065.

Paulraj, A. J., Gore, D. A., Nabar, R. U., and Bolcskei, H. 2004. An overview of MIMO communications: A key to gigabit wireless. *Proceedings of the IEEE* 92(2):198–218.

Raab, F. H., Asbeck, P., Cripps, S., Kenington, P. B., Popovic, Z. B., Pothecary, N., Sevic, J. F., and Sokal, N. O. 2002. Power amplifiers and transmitters for RF and microwave. *IEEE Transactions on Microwave Theory and Techniques* 50(3):814–826.

Rajbanshi, R., and Veeragandham, A. 2004. OFDM systems design. Doctoral dissertation, Information and Telecommunication Technology Center (ITTC), University of Kansas.

Roca, A. 2007. Implementation of a WiMAX simulator in Simulink. Graduate engineer dissertation, Technical University of Vienna, Spain.

Sen, S., Senguttuvan, R., and Chatterjee, A. 2008. Concurrent PAR and power amplifier adaptation for power efficient operation of WiMAX OFDM transmitters. Proceedings of IEEE Radio and Wireless Symposium (RWS), Orlando, FL, January 21–24.

Stewart, B. G., and Vallavaraj, A. 2009. BER performance evaluation of tail-biting convolution coding applied to companded QPSK mobile WiMax. Proceedings of 15th International Conference on Parallel and Distributed Systems (ICPADS), Shenzhen, China, December 8–11.

Visser, M. A., and Bar-Ness, Y. 1999. Frequency offset correction for OFDM using a blind adaptive decorrelator in a time-variant selective Rayleigh fading channel. Proceedings of 49th IEEE Vehicular Technology Conference, Houston, TX, May 16–20.

Zhang, L., and Thibault, L. 2010. Performance evaluation of mobile DAB receivers in enhanced packet mode. *IEEE Transactions on Consumer Electronics* 56(4):2115–2122.

3

POWER AMPLIFIERS IN WIRELESS COMMUNICATIONS

3.1 Introduction

Power amplifiers (PAs) are some of the important and costly devices in communication systems and they are nonlinear in a certain operating region. The nonlinearity of the PA has several undesired impacts, such as spectral broadening that causes adjacent channel interference (ACI) and in-band distortion that causes phase distortion, which is measured by error vector magnitude (EVM). There are several ways to tackle the nonlinearity problem of PAs. One simple solution is to back off the PA from its saturation point to force it to operate within its linear region. But in this way a significant loss in power efficiency will occur, typically less than 10% (Wright, 2002), and more than 90% of the direct current (DC) power is lost and turns into heat. By increasing the number of the base stations and then the number of power amplifiers improves the efficiency of the power amplifier and reduces the cost of the system. Moreover, transmission formats (e.g., code division multiple access [CDMA] and orthogonal frequency division multiplexing [OFDM]) suffer from high peak-to-average power ratios (PAPRs) or the crest factor (CF), that is, large fluctuations of signal envelopes. To improve the power amplifier efficiency without compromising its linearity, power amplifier linearization is essential.

Another important fact is that with the increasing number of users, a greater amount of bandwidth is required. One effective way to increase bandwidth is to use diversity techniques, which have been applied in most of the third-generation (3G) standard specifications (Vuolevi, 2003). But, with each additional antenna, an additional transceiver is required, which can significantly increase the system cost. Another method is the digital predistortion (DP) technique. The DP technique overcomes the linearity problem of PAs, enabling the use of nonlinear PAs that are cheaper, thereby reducing the cost of the overall system (Kenington, 2000). Digital predistortion among all linearization techniques is the one that is low in cost, one with high efficiency and also high flexibility. By applying digital predistortion, which is implemented in the baseband of the communications system, the nonlinearity of the power amplifier is reduced and it allows the use of a high power amplifier with high efficiency in the system.

Another important fact in the study of PAs is memory effects, which are the main issue of this book. The focus here is on short-term memory effects, which cause the characteristics of PAs to vary with time. This effect is more important when high bandwidth signals are applied. The memory effects cause an increase in adjacent channel leakage ratio (ACLR) and also error vector magnitude (EVM), which were explained in Chapter 1. The other factor that might cause problems to the performance of the predistortion is the effect of the noise. Noise here can be the result of the analog part such as the digital-to-analog converter (DAC), mixer, and so on, which in this book are not considered, because the predistortion here is implemented in baseband. The other noise source also can be from the PA, which will be under memory effects and with the feedback that is in the adaptive predistortion that noise will be cancelled. In this book the linearization techniques of PAs, mainly class AB, are investigated, then the digital predistortion technique is chosen since it is the most cost effective and most efficient among all the linearization techniques. The class AB power amplifier is chosen because it has more linearity as compared to other classes. A new technique has been developed, simulated, and experimentally measured to validate this new technique. Finally, the simulation and experimental results are compared.

This chapter begins with the characteristics of the power amplifier. The nonlinearity of the power amplifier will be discussed and analyzed in further detail. The memory effects of the power amplifier will also be briefly discussed for the purpose of perfecting the implementation effort. The nonlinearity of the power amplifier causes low efficiency if the transmitting power is forced to back off too far from the power amplifier's saturation point. Linearizing the power amplifier creates various benefits especially to network and telecommunications companies. This is because an efficient power amplifier in the transmitter increases the performance of the overall communication system.

3.2 High Power Amplifiers

A power amplifier is used to amplify input signals into output signals with a larger amount of power and higher amplitude. The amplifying process is needed so that both the transmitting and receiving parties can communicate successfully within the desired distance. The key parameters of a power amplifier include an achievable output power level, linearity, and efficiency. There are many ways to define the efficiency of the power amplifier but two basic ones include:

1. *Drain efficiency*—The ratio between the radio frequency (RF) output power to the DC consumed power.
2. *Power-added efficiency (PAE)*—The ratio of the difference between radio frequency (RF) output power and the input power to the DC consumed power.

The PAE takes into account the gain of the power amplifier. More stages are required to increase the power when the power gain decreases. In each stage, a certain amount of power is consumed, which in turn adds up to the overall power consumption. Low gain of the power amplifier therefore increases the overall power consumption, thus decreasing the overall efficiency (Kenington, 2000).

3.2.1 Nonlinearity of Power Amplifiers

There are three regions for a power amplifier with amplitude dependent characteristics (AM-AM).

1. *Cutoff region*—The region where the amplifier is not conducting any amplification.
2. *Linear region*—The region where the amplifier starts amplifying the input signal.
3. *Saturation region*—The region where the amplifier starts to exhibit nonlinearity, causing the output to be saturated and remain unchanged although the input power increases.

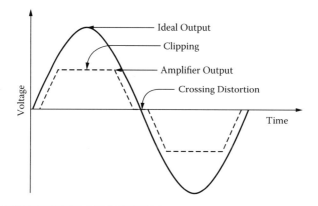

Figure 3.1 Power amplifier distortion characteristics. (From Patel, J. 2004. Adaptive digital predistortion linearizer for power amplifiers in a military UHF satellite. Department of Electrical Engineering, University of South Florida, March. With permission.)

Figure 3.1 compares the ideal amplifier output and the actual amplifier output, which has undergone clipping when the amplifier is operating in the saturation region. The voltage level shown here is the maximum voltage output that could be amplified by the power amplifier. Gain compression happens until maximum power is achieved. This power amplifier saturation also causes phase modulation of the signal.

The main characteristics of a power amplifier are as follows:

- *Second-order intercept point*—The point where the second harmonic signal clashes with the first-order ideal linear response. The second harmonic signal increases in reference to the input signal by the order of two.
- *Third-order intercept point*—The point where the third harmonic signal clashes with the first-order ideal linear response. The third harmonic signal increases in reference to the input signal by the order of three.
- *1 dB gain compression point*—The point where the actual power level is 1 dB lower than the ideal power level, caused by the nonlinearity of the power amplifier.
- *Input backoff (IBO)*—The ratio of the input power measured at the power amplifier to the input signal power that produces the maximum output signal power.

The graph in Figure 3.2 shows the corresponding interception points together with the 1 dB compression point.

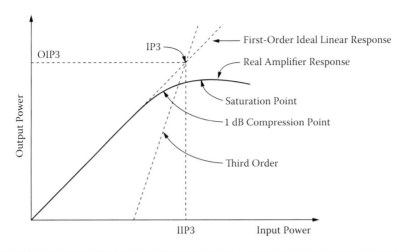

Figure 3.2 Compression and intercept points of power amplifiers. (From Varahram, P. et al. 2009. *Journal of Electrical Engineering* 60(3):129–135. With permission.)

3.3 Characteristics of Power Amplifiers

The linearity of a power amplifier was discussed earlier in this chapter. In this chapter the other characteristics of a power amplifier, including efficiency, will be discussed.

It should be noted that there is a trade-off between linearity and efficiency: the higher the linearity of a power amplifier, the lower its efficiency. Hence, the power amplifier has to be designed to address these challenges. It is important to know the application for which the power amplifier is going to be used.

3.3.1 Efficiency

The amplifier's efficiency is defined as a measure of how effectively the power amplifier can convert the DC power of the supply into the signal power to be delivered to the load, expressed as follows:

$$\eta = \frac{Signal\ Power\ Delivered\ to\ Load}{DC\ Power\ Supplied\ to\ Output\ Circuit} = \frac{P_{out}}{P_{DC}} \tag{3.1}$$

where P_{out} is the output RF power and P_{DC} is the power taken from the DC source.

In other words, efficiency is a measure of how well a device converts one energy source to another. In microwave engineering, we are interested in converting DC power to RF power. So, efficiency in power amplifiers describes the part of the DC power that is converted to RF power.

The efficiency is one for an ideal power amplifier; this means that the power delivered to the load is equal to the power derived from the DC supply, which states that there is no power consumption in the power amplifier. However, in reality this will not be happening, due to the fact that there is no fully linear power amplifier in real applications.

The following section describes how amplifier efficiency can be calculated.

3.3.1.1 Drain Efficiency

As shown in Equation (3.2), drain efficiency is defined as the ratio of RF output, $P_{RF_{out}}$, to DC input power, P_{DC}, so rises in roughly the same proportion as the fundamental output power if the DC power is constant.

$$\delta_{drain} = \frac{P_{RF_{out}}}{P_{DC}} = \frac{P_{RF_{out}}}{V_{DC} \times I_{DC}} \qquad (3.2)$$

Drain efficiency as expressed by Equation (3.2) indicates how much DC power can be converted to RF power. The only problem with this definition is that it does not consider the RF power that goes into a device.

3.3.1.2 Power-Added Efficiency (PAE)

Power-added efficiency is similar to drain efficiency, but it takes into account the RF power that is added to the device at its input. PAE is defined as

$$PAE = \frac{(P_{out} - P_{in})}{P_{dc}} \qquad (3.3)$$

3.3.2 Output Power

The output power has an important role in order to design a power amplifier. To measure and evaluate the output power, a parameter, P_{MAX}, is used that indicates the output power when 1 volt and 1 amp are on the drain of the field effect transistor (FET). Multiplication of P_{MAX} by the drain voltage and current of a device produces the maximum output power available from that device.

The power output factor is defined by

$$P_{MAX} = \frac{The\ Maximum\ Out\ Power}{The\ Peak\ Drain\ Voltage\ \times\ The\ Peak\ Drain\ Current} \qquad (3.4)$$

3.3.3 Signal Gain

The gain of the amplifier (G) is defined as the magnitude of the output signal (X_o) over the magnitude of the input signal (X_i) and can be expressed as follows:

$$G = \frac{X_o}{X_i} \qquad (3.5)$$

where G is the voltage, current, or power gain depending on the application.

3.3.4 Trade-Off between Linearity and Efficiency

The power amplifier efficiency in communication systems is one of the most important factors for prolonging battery life and saving the cost of the systems. In base stations, power efficiency is also very important, since it has an impact in both power usage and cooling requirements (Cripps, 2006). Power amplifiers are typically the most power-consuming components in communication systems and can represent up to 70% of the total power consumption in a subsystem of communication systems (O'Droma et al., 2004).

The efficiency of an amplifier is a measure of how effectively DC power is converted to RF power and can be expressed as (Kenington, 2000)

$$h = \frac{P_{out}}{P_{DC}}\ \% \qquad (3.6)$$

In Equation (3.6), P_{out} is the output power of the power amplifier and P_{DC} is the DC power of the power amplifier. But, by just considering the DC power converted into RF power, the measure does not have to consider the power that already exists in RF power and is injected into the power amplifier. Therefore, the PAE is defined as (Kenington, 2000)

$$PAE = \frac{P_{out} - P_{in}}{P_{DC}} \%$$

(3.7)

In modern spectrally efficient communication systems, modulation with nonconstant envelope signals is used. These modulation schemes produce signals with high PAPRs. The need to avoid nonlinear effects in the amplification requires the transmission of signals with peak amplitudes well below the output peak of the amplifier (backoff operation), which degrades the average efficiency (Vuolevi, 2003). The power amplifier efficiency not only depends on the input or output backoff chosen to operate the PA, but also depends on the power transistor operation class.

Figure 3.3 shows some of the most common power amplifier operation classes, which are distinguished by the fraction of the RF cycle over which the power transistor conducts. This means that theoretically 100% corresponds to class A; 50% to class B; between 50% and 100% to class AB; and less than 50% to classes C, D, E, and F.

The trade-off between the different operation classes includes efficiency, linearity, power gain, signal bandwidth, and output power (Raab et al., 2002). Table 3.1 shows the trade-off between linearity and efficiency according to the power transistor's operation class.

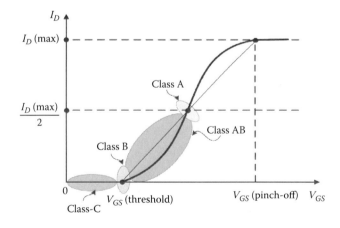

Figure 3.3 Power amplifier classes. (From Saleh, A. A. M. 1981. *IEEE Transactions on Communications* 29:1715–1720, November. With permission.)

Table 3.1   Efficiency and Linearity of Power Amplifier Classes of Operation

CLASS OF OPERATION	OPERATION MODE	MAXIMUM EFFICIENCY	LINEARITY
Class A	Current source mode	50	Good
Class AB	Current source mode	Better than class A, worse than class B	Better than class B, worse than class A
Class B	Current source mode	78.5	Average
Class C	Current source mode	100	Poor
Class D	Switch mode	100	Poor
Class E	Switch mode	100	Poor
Class F	Switch mode	100	Poor

Source: Cripps, S. C. 2006. *RF Power Amplifiers for Wireless Communications*, Norwood, MA: Artech House.

Whereas in class A, class AB, class B, and class C the transistor behaves as a transconductor, in class D, class E, and class F it behaves as a switch and has high-efficiency levels. Although class C, class D, class E, and class F power amplifiers have high efficiency, they are often not suitable for linear applications because of the output distortions that these classes of power amplifiers creates. Normally, these classes of power amplifiers use constant envelope signals. However, by using a suitable linearizing circuit, these efficient power amplifiers have been used for modulation schemes with linear amplification (Gupta et al., 2001).

Therefore, the use of class A power amplifiers with backoff ensures linear amplification but with loss of efficiency. On the other hand, the use of a more efficient class of power amplifiers, such as class C or AB, the efficiency is more but the less nonlinearity causes the need of a more accurate linearization technique.

For an understanding of each class of power amplifier, see Section 3.4.

3.3.5 Power Amplifier Two-Tone Test

A two-tone test is used to measure the amplitude and phase distortions in a power amplifier. In the two-tone test, the envelope of the input signal is changed throughout its complete range, then the amplifier is tested over its whole transfer characteristics. The input signal can be expressed by (Patel, 2004)

$$V_{in}(t) = b\sin(\omega_1 t) + b\sin(\omega_2 t) \tag{3.8}$$

Then the output voltage is (Patel, 2004)

$$V_o(t) = a_1 b[\sin(\omega_1 t) + \sin(\omega_2 t)] + a_2 b^2 [\sin(\omega_1 t) + \sin(\omega_2 t)]^2$$
$$+ a_3 b^3 [\sin(\omega_1 t) + \sin(\omega_2 t)]^3 + a_4 b^4 [\sin(\omega_1 t) + \sin(\omega_2 t)]^4$$

(3.9)

Each product term in Equation (3.9) generates distortion products. But the even order intermodulation distortion (IMD) terms are out of the bandwidth of the signal, so they are not considered. The odd order IMD terms are more important because they are in the bandwidth of the signal as shown in Figure 3.4. As shown in Figure 3.4, the IMD distortion causes major problems in a communication system as opposed to harmonic distortion.

A two-tone test represents an approximation to characterize the nonlinear power amplifier behavior. Normally, the signals passing through an amplifier are modulated signals characterized by complex frequency spectra. When complex modulation signals pass through the power amplifier that work near the saturation point, nonlinearities appear over a continuous band of frequencies and this is referred to as the spectral regrowth. Figure 3.5 shows the spectra of both input and amplified output of a two-carrier WCDMA (wideband code division multiple access) modulated signal. The output spectrum shows spectral regrowth due to the nonlinear behavior of the power amplifier (Pinal, 2007).

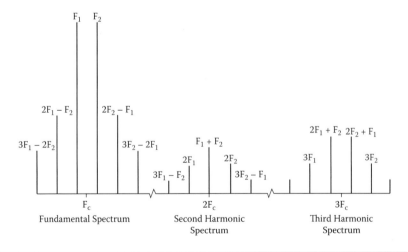

Figure 3.4   Harmonic distortion of the two-tone test. (From Patel, J. 2004. Adaptive digital predistortion linearizer for power amplifiers in a military UHF satellite. Department of Electrical Engineering, University of South Florida, March. With permission.)

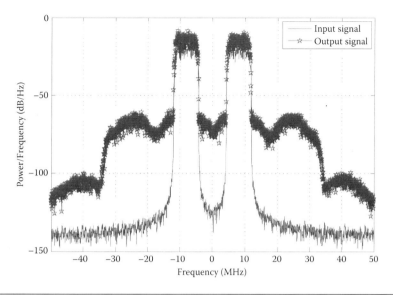

Figure 3.5   Input and output spectrum of a power amplifier for a two-carrier WCDMA modulated signal.

The parameter for measuring the amount of the out-of-band distortion of the power amplifier is the adjacent channel power ratio (ACPR) or also known as the adjacent channel leakage ratio (ACLR), and it is defined as the ratio of the total power over the channel bandwidth to the power delivered in the adjacent channels (both upper sideband [US] and lower sideband [LS]).

The following equation shows the formula for calculating ACPR (Pinal, 2007):

$$ACPR = \frac{P_{in-band}}{P_{adjacent-channel}} = \frac{\int\limits_{B} P_{out}(f).df}{\int\limits_{LS} P_{out}(f).df + \int\limits_{US} P_{out}(f).df}[dBr] \quad (3.10)$$

In Equation (3.10), $P_{in-band}$ is the total power in the bandwidth and $P_{adjacent-channel}$ is the total power in the adjacent channel.

ACLR indicates the minimum adjacent power level required to not interfere with the adjacent channels. It is then necessary to avoid high levels of spectral regrowth. This objective can be easily achieved by operating at a very linear region of the PA, that is, with increasing the IBO, but this will also cause the efficiency of the power amplifier to drop significantly.

Nonlinear distortion in power amplifiers can cause changes in nonconstant envelope signals, which modulate amplitude and phase (or I and Q) together. Therefore, modulation formats like quadrature amplitude modulation (QAM) can suffer from nonlinear distortion, which can be measured by the EVM. EVM is an effective method for both characterizing signal distortion and measuring overall performance of communication systems (Heutmaker, 1997; Voelker, 1995).

Figure 3.6 graphically shows the error vector between the desired (reference) signal and the measured signal. The EVM is defined as the square root of the ratio of the mean error vector power to the mean reference power (Kenington, 2000):

$$EVM = \sqrt{\frac{\dfrac{1}{N}\displaystyle\sum_{1}^{N}(DI^2 + DQ^2)}{S_{max}^2}}\,[\%] \qquad (3.11)$$

The EVM includes information on the filter accuracy of the transmitter, DAC, modulator imbalances, and power amplifier nonlinearity (Kenington, 2000).

3.4 Classification of Power Amplifiers

In this section, the classification of the power amplifiers based on their circuit and operation will be studied. In general, power amplifiers can be classified in different classes: A, B, AB, C, D, E, and so on.

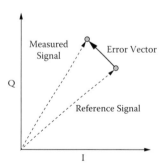

Figure 3.6   Error vector representation. (From Pinal, P. L. G. 2007. Multi look-up table digital predistortion for RF power amplifier linearization. Department of Signal Theory and Communications, Universitat Politecnica de Catalunya, Barcelona, Spain. With permission.)

In other classifications, a power amplifier is categorized based on the linearity and efficiency. As mentioned earlier, the higher the linearity of the power amplifier, the lower its efficiency. Most of this information and details can be found in the book by Krauss, Bostian, and Raab, *Solid State Radio Engineering* (1980).

The parameters of FET that are being assumed to calculate the efficiencies are

$$
\left[
\begin{array}{lll}
i_D = 0 & Cutoff & Region \\
i_D = g_m \cdot (V_{GS} - V_T) & Active & Region \\
i_D = \dfrac{V_D}{R_{on}} & Saturation & Region
\end{array}
\right.
\qquad (3.12)
$$

The regions of operation are defined by

$$
\begin{array}{lll}
Cutoff\ region & V_{GS} < V_T & \\
Active\ region & V_{GS} \geq V_T & and \quad i_D < \left(\dfrac{V_D}{R_{on}}\right) \\
Saturation\ region & V_{GS} \geq V_T & and \quad i_D = \left(\dfrac{V_D}{R_{on}}\right)
\end{array}
\qquad (3.13)
$$

where "saturation" means the maximum point that the power amplifier can generate power and beyond that the power amplifier will be burnt.

3.4.1 Class A

Among all classes of power amplifiers, class A has the highest linearity. This is due to the linear operation of this class of power amplifier, which is basically equivalent to a current source. Figures 3.7 and 3.8 show the configurations of class A, B, and C power amplifiers that are in the form of a push–pull or a single-ended tune. In Figure 3.9, the load line and current waveform of the class A amplifier is shown. To achieve high linearity and gain, the amplifier's base and drain DC voltage should be chosen properly so that the amplifier operates in the linear region.

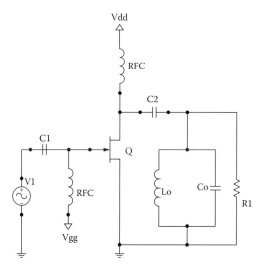

Figure 3.7 Single-ended power amplifier (class A, B, or C). (From Pinal, P. L. G. 2007. Multi look-up table digital predistortion for RF power amplifier linearization. Department of Signal Theory and Communications, Universitat Politecnica de Catalunya, Barcelona, Spain. With permission.)

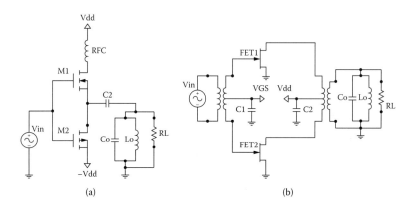

Figure 3.8 (a) Push–pull power amplifier (class A, B, or C). (b) Transformer-coupled push–pull power amplifier (class A, B, or C). (From Pinal, P. L. G. 2007. Multi look-up table digital predistortion for RF power amplifier linearization. Department of Signal Theory and Communications, Universitat Politecnica de Catalunya, Barcelona, Spain. With permission.)

As shown in Figure 3.7, the maximum AC (alternating current) output voltage V_{om} is slightly less than V_{DD} and the maximum AC output current I_{om} is equal to I_{dq}. The drain voltage must have a DC component equal to that of the supply voltage and a fundamental-frequency component equal to that of the output voltage; hence

$$V_D(\theta) = V_{DD} + V_{om} \cdot \sin\theta \qquad (3.14)$$

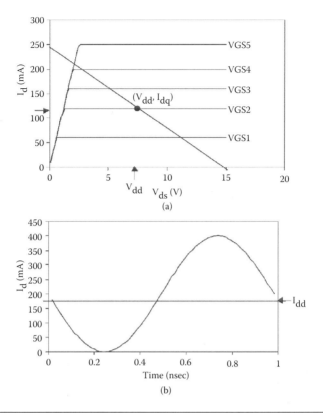

Figure 3.9   (a) Load line and (b) current waveform for the class A power amplifier. (From Pinal, P. L. G. 2007. Multi look-up table digital predistortion for RF power amplifier linearization. Department of Signal Theory and Communications, Universitat Politecnica de Catalunya, Barcelona, Spain. With permission.)

The DC power is

$$P_{dc} = V_{DD} \cdot I_{dq} \tag{3.15}$$

The maximum output power is

$$P_o = \frac{1}{2} \cdot V_{om} \cdot I_{om} \approx \frac{1}{2} \cdot V_{DD} \cdot I_{dq} \tag{3.16}$$

And the efficiency is

$$\eta = \frac{P_o}{P_{dc}} \cdot 100 = \frac{1}{2} \cdot \frac{V_{om}}{V_{DD}} \cdot 100 \le 50\% \tag{3.17}$$

The power dissipation is a difference between the DC power and output power and can be expressed as follows:

$$P_d = P_{dc} - P_o \tag{3.18}$$

3.4.2 Class B

The operation of a class B power amplifier is different than a class A in that it operates at a zero quiescent current, hence the DC power is small. This causes its efficiency to be higher than a class A power amplifier. This comes at the expense of the linearity of the power amplifier.

Figure 3.10 shows how the class B amplifier operates. The output power of the single-ended class B amplifier is

$$P_o = \frac{1}{2} \cdot I_o \cdot V_o \qquad (3.19)$$

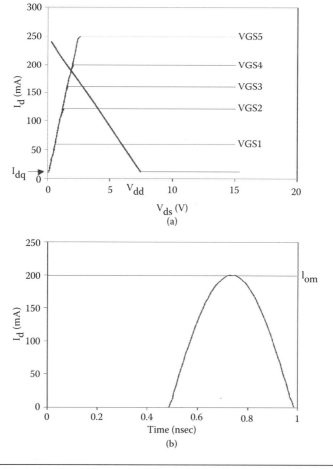

Figure 3.10   (a) Load line and (b) current waveform for the class B power amplifier. (From Pinal, P. L. G. 2007. Multi look-up table digital predistortion for RF power amplifier linearization. Department of Signal Theory and Communications, Universitat Politecnica de Catalunya, Barcelona, Spain. With permission.)

The DC drain current is

$$I_{dc} = 2\frac{I_{om}}{\pi} \tag{3.20}$$

The DC power is

$$P_{dc} = 2\frac{I_{om} \cdot V_{DD}}{\pi} \tag{3.21}$$

and the maximum efficiency when $V_{om} = V_{DD}$ is

$$\eta = \frac{P_o}{P_{dc}} \cdot 100 = \frac{\pi}{4} \cdot \frac{V_{om}}{V_{DD}} \cdot 100 \le 78.53\% \tag{3.22}$$

3.4.3 Class AB

The other class of power amplifier, which is a compromise between class A and class B, is class AB. This class of power amplifiers is mostly used in mobile and wireless communications systems as a solid-state power amplifier (SSPA) due to its trade-off between efficiency and linearity.

3.4.4 Class C

The previous classes—A, B, and AB—are considered as linear amplifiers, where the input and output signals are linear to each other. In some applications, where linearity is not an issue and efficiency is critical, non-linear amplifier classes (C, D, E, or F) are used.

Figure 3.11 illustrates the operation of the class C amplifier.

Class A, AB, B, and C amplifiers can be defined as follows:

$$\text{Class of Operation} = \begin{cases} A, & y = \pi \\ B, & y = \frac{\pi}{2} \\ AB, & \frac{\pi}{2} \langle y \langle \pi \\ C, & y \langle \frac{\pi}{2} \end{cases} \tag{3.23}$$

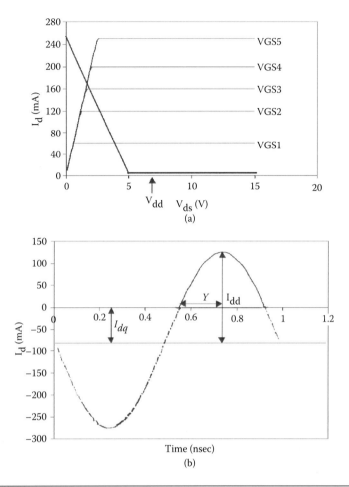

Figure 3.11 (a) Load line and (b) current waveform for the class C power. (From Pinal, P. L. G. 2007. Multi look-up table digital predistortion for RF power amplifier linearization. Department of Signal Theory and Communications, Universitat Politecnica de Catalunya, Barcelona, Spain. With permission.)

The conduction angle is

$$Y = \arccos\left(-\frac{I_{dq}}{I_{dd}}\right) \tag{3.24}$$

The DC current is

$$I_{dc} = \frac{1}{2\pi} \cdot \int_{0}^{2\pi} i_D(\theta)\, d\theta = \frac{1}{\pi} \cdot \left(\begin{array}{l} I_{dq} \cdot y - I_{dd} \cdot \sin(y) = \dfrac{I_{dd}}{\pi} \\[2mm] \cdot (\sin(y) - y\cos(y)) \end{array} \right) \tag{3.25}$$

The output power is

$$P_o = \frac{V_o^2}{R} \tag{3.26}$$

The DC power is

$$P_{dc} = V_{cc} \cdot I_{dd} \tag{3.27}$$

and the maximum output voltage, V_o, is

$$V_{OMAX} = V_{DD} \tag{3.28}$$

From the preceding equations, the maximum efficiency is

$$\eta_{max} = \frac{P_{OMAX}}{P_i} = \frac{2y - \sin(2y)}{4 \cdot [\sin(y) - y \cdot \cos(y)]} \tag{3.29}$$

The peak drain voltage and drain current are

$$V_{DMAX} = 2\,V_{DD} \tag{3.30}$$

and

$$I_{DMAX} = I_{dq} + I_{dd} \tag{3.31}$$

The power output capability factor is

$$\eta_{max} = \frac{P_{OMAX}}{V_{DMAX} \cdot I_{DMAX}} = \frac{2y - \sin(2y)}{8\pi \cdot [1 - \cos(y)]} \tag{3.32}$$

Figure 3.12 shows the efficiency versus the conduction angle. It shows that 100% efficiency is possible, however, it is impractical because the output power is zero, as shown in Figure 3.13.

Figure 3.14 illustrates the schematic and current waveforms for the aforementioned classes of operation.

3.4.5 Class F

The class F amplifier has the highest efficiency among the other classes of amplifiers. It uses harmonic resonators to achieve high efficiency, which results from a low DC voltage current product. Figure 3.15 shows a class F amplifier.

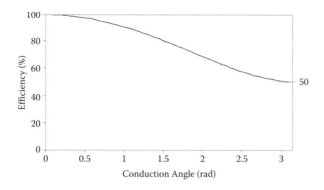

Figure 3.12 Efficiency versus conduction angle.

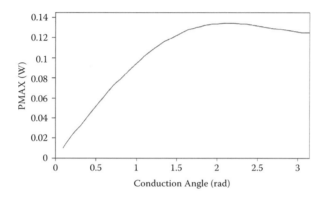

Figure 3.13 P_{MAX} versus conduction angle.

The drain voltage is

$$V_d(\theta) = V_{DD} + V_{om} \cdot \sin\theta + V_{om3} \cdot \sin(3\theta) \tag{3.33}$$

The setting $V_{om3} = \dfrac{V_{om}}{9}$ produces maximum flatness for the drain voltage. And, the maximum output occurs when the minimum point of $V_d(\theta)$ is zero. Hence,

$$V_{om} = \frac{9}{8} \cdot V_{DD} \tag{3.34}$$

The DC current is

$$I_{dc} = \frac{I_{dm}}{\pi} \tag{3.35}$$

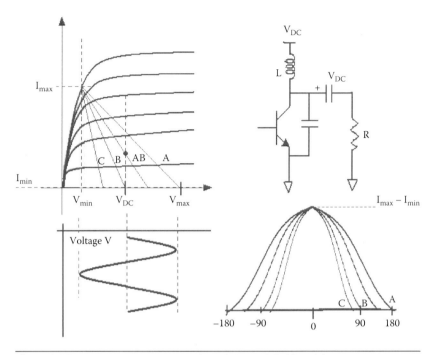

Figure 3.14 Schematic and current waveform of classes A, B, AB, and C.

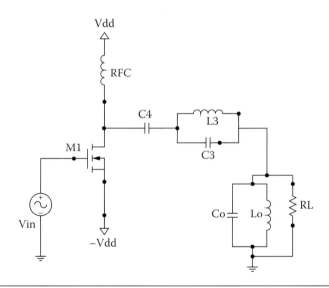

Figure 3.15 Single-ended power amplifier (class F). (From Pinal, P. L. G. 2007. Multi look-up table digital predistortion for RF power amplifier linearization. Department of Signal Theory and Communications, Universitat Politecnica de Catalunya, Barcelona, Spain. With permission.)

The DC power is

$$P_{dc} = V_{DD} \cdot \frac{I_{dm}}{\pi} \tag{3.36}$$

The fundamental current is

$$I_{om} = \frac{I_{dm}}{2} \cdot \sin \theta \tag{3.37}$$

and the maximum efficiency is

$$\eta_{max} = \frac{P_{o\ max}}{P_{dc}} \cdot 100 = \frac{\dfrac{I_{dm}}{4} \cdot \dfrac{9}{8} V_{DD}}{\dfrac{I_{dm}}{\pi}} \cdot 100 = 88.36\% \tag{3.38}$$

3.4.6 Other High-Efficiency Classes

There are other high-efficiency amplifiers, including D, E, G, H, and S. These classes use different techniques to reduce the average collector or drain power that increase the efficiency. Classes D, E, and S use a switching technique, whereas classes G and H use resonators and multiple power-supply voltages to reduce the collector current-voltage product.

3.5 Power Amplifier Memory Effects

In this book, the power amplifiers are modeled with memory effects and later this model is shown in Section 4.5.3.. The memory effect is characterized by the fluctuation of the frequency domain transfer function of the power amplifier or equivalently, the time dependence of the transfer function. The effect of memory on the PA output can be shown with a frequency domain plot of the power amplifier output, which is illustrated in Figure 3.16. IM_L and IM_U are the intermodulation results. The height of the intermodulation distortions shows the power of the components, and the angles ϕ_L and ϕ_U represent the phase shift of the intermodulation results. The power of the intermodulation products and the phase shift are not exactly the same, and can also change depending on

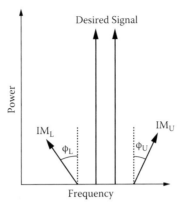

Figure 3.16 Principle of distortion cancellation and its sensitivity to memory effects. (From Vuolevi, J. 2003. *Distortion in RF Power Amplifiers,* Norwood, MA: Artech House. With permission.)

the frequency separation of the two-tone signals. This causes problems with the memoryless linearization method, since it tries to compensate upper and lower intermodulation products similarly, and then at least one of them should be compensated. This shows that upper and lower intermodulation distortions are reduced differently.

The dependence of the intermodulation distortion components on the separation of the two-tone signals can be used to define the memory effects in a power amplifier (Kenington, 2000) by testing the power amplifier with two-tone signals with different spacing on the wanted signal band and noting the behavior of the intermodulation results. However, modeling a power amplifier with memory requires more complicated measurements, which are described in Ku and Kenney (2003) and Vuolevi et al. (2001).

There are two types of memory effects, electrical and thermal, which are discussed next.

3.5.1 Electrical Memory Effects

Electrical memory effects (also known as short-term memory effects) are mainly caused by nonconstant terminal impedances at the DC, fundamental and harmonic bands, and due to envelope impedances. The Volterra series can model this type of memory effects. The main focus of this book is on this type of memory effect, which we will investigate to overcome this impact.

3.5.2 Electrothermal Memory Effects

Electrothermal memory effects (also called long-term effects) are caused by temperature variations at the top of the power amplifier and heat sink. The result is that third-order intermodulation distortion (IMD3) signals are generated. This type of memory effect normally appears at low bandwidth applications below 1 MHz, because the mass of the semiconductor in the active device cannot change its temperature fast enough to keep up with high envelope frequencies. Among the different linearization techniques used for broadband transmitters, digital predistortion is the most sensitive to memory effects. In order to cancel out or minimize memory effects, three techniques are considered in the literature: impedance optimization, envelope injection, and envelope filtering. Impedance optimization and envelope injection are both at the baseband bias impedances. Impedance optimization is based on the optimization of the out-of-band impedances. While in the envelope injection technique, a low-frequency envelope signal is generated and added to the RF carrier. For further details, these two techniques can be found in Vuolevi (2003). Finally, in the envelope filtering technique, the objective is to create inverse memory effects that are generated inside the PA. Therefore, the digital predistorter not only has to compensate the PA's nonlinear behavior but also has to compensate the memory effects, which is done by filtering and phase shifting the envelope signal.

3.5.3 Modeling Power Amplifiers

Before introducing and studying the linearization techniques it is important to be familiar with power amplifier modeling. Depending on the accuracy, there are different power amplifier models. The memoryless power amplifier model is discussed first, and then the power amplifier with memory, which is used in this book for modeling a power amplifier, is presented.

3.5.4 Modeling Power Amplifiers without Memory

The memoryless power amplifier can be modeled as a nonlinear function that maps a real-valued input to a real-valued output.

This memoryless nonlinearity can be approximated by a power series as (Ding, 2004):

$$\tilde{y}(t) = \sum_{k=1}^{K} \tilde{a}_k \tilde{x}^k(t) \tag{3.39}$$

where $\tilde{a}_k$ are real-valued coefficients, $\tilde{x}(t)$ is the passband power amplifier input, and $\tilde{y}(t)$ is the passband power amplifier output. In the baseband, Equation (3.39) becomes (Ding, 2004)

$$y(t) = \sum_{\substack{k=1 \\ odd}}^{K} a_k x(t) |x(t)|^{k-1} \tag{3.40}$$

where

$$a_k = 2^{1-k} \binom{k}{\frac{k-1}{2}} \tilde{a}_k$$

$x(t)$ is the baseband power amplifier input, and $y(t)$ is the baseband power amplifier output.

Note that Equation (3.40) only contains odd order terms. This is because the signals generated from the even order terms in Equation (3.39) are far from the carrier frequency, so they do not affect the baseband output $y(t)$. It is also important to note that the coefficient a_k is real valued since $\tilde{a}_k$ is real valued. This means that if the power amplifier is memoryless, it only has amplitude distortion of the input signal, which is the AM-AM conversion of the power amplifier. So, the power amplifier only has amplitude distortion and does not have phase distortion.

Now consider if $\tilde{a}_k$ in Equation (3.40) is complex, which represents a much larger class of power amplifiers, which are referred to as quasi-memoryless power amplifiers. In the passband, a nonlinear power amplifier with memory can be approximated by the Volterra series as follows (Ding, 2004):

$$\tilde{y}(t) = \sum_{K} \int \cdots \int \tilde{v}_k(t_k) \prod_{i=1}^{k} \tilde{x}(t - t_i) dt_k \tag{3.41}$$

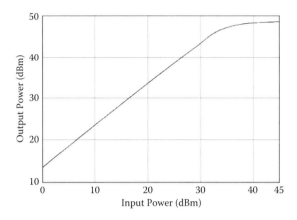

Figure 3.17   The AM-AM and AM-PM responses of a class AB power amplifier.

where $\tau_k = [\tau_1, ... \tau_k]^T$, $\tilde{v}_k(.)$ is the real-valued kth-order Volterra kernel, and $d\tau_k = d\tau_1 \, ... \, d\tau_k$. A special case of the Volterra series is when the power amplifier has short-term memory; on the other hand, the time span of the memory is short compared to the time variations of the input signal envelope. Raich and Zhou (2002) have shown that the baseband version of Equation (3.40) has the same form as Equation (3.41) except that a_k are complex valued. Now the power amplifier has both amplitude distortion and phase distortion to the input signal. However, the baseband representation in this case is still memoryless according to Equation (3.39). The concentration of this book is on the power amplifier model with memory, which is explained in the following section. As an example, the AM-AM and AM-PM responses of a memoryless class AB power amplifier are shown in Figure 3.17. As seen in this figure, the power amplifier characteristics do not change with time and this is exactly the case with the quasi-memoryless power amplifier.

3.5.5 Power Amplifier Model with Memory Effects

In Section 3.5.4, the passband and baseband representation of the power amplifier without memory effects was discussed. Also, as shown in Appendix A, the passband signal can be represented with a baseband model. As the input signal bandwidth increases, such as in WCDMA, the time span of the power amplifier memory becomes comparable to the time variations of the input signal envelope. Therefore, the memory effects of the power amplifier are not considered as short term.

Then, Equation (3.42) gives the full baseband Volterra series (Benedetto et al., 1979; Raich and Zhou, 2002):

$$y(t) = \sum_K \int \cdots \int h_{2k+1}(\tau_{2k+1}) \prod_{i=1}^{k+1} x(t-\tau_i) \prod_{i=k+2}^{2k+1} x^*(t-\tau_i) d\tau_{2k+1} \quad (3.42)$$

where

$$h_{2k+1}(t_{2k+1}) = \frac{1}{2^{2k}} \binom{2k+1}{k} \tilde{v}_{2k+1}(t_{2k+1}) e^{-2pf_o \left(\sum_{i=1}^{k+1} t_i - \sum_{i=k+2}^{2k+1} t_i \right)} \quad (3.43)$$

From Equation (3.43), it can be seen that the number of coefficients of the Volterra series increases exponentially as the memory length and the nonlinear order increase. This drawback makes the Volterra series impractical for real-time applications. Therefore, several special cases of the Volterra series should be studied. The special cases considered here include the Wiener model, the Hammerstein model, the Wiener-Hammerstein model (Benedetto et al., 1983), the memory polynomial model, and the Murray Hill model (Ding, 2004). Next, the memory polynomial model is explained in detail.

The memory polynomial model is the best approximation of the Volterra series, which uses the diagonal kernels of the Volterra series. In the discrete-time Volterra series, Equation (3.43), if $l_1 = \ldots = l_{2k+1} = l$, becomes

$$y(n) = \sum_{\substack{p=1 \\ odd}}^{P} \sum_{l=0}^{L-1} a_{pl} x(n-l) |x(n-l)|^{p-1} \quad (3.44)$$

where a_{pl} is equal to $v_{2k+1}(l,l,\ldots,l)$ in Equation (3.44). This model was considered for modeling power amplifiers with memory effects by Kim (2001), and for data predistortion of the cascade of a pulse shaping filter and a memoryless power amplifier by Chang et al. (2001). In this book, we use this model for comparison between the new predistortion technique that is proposed here and memory polynomial for power amplifiers.

In another research study of power amplifiers with memory effects, which was first introduced by Ku and Kenney (2003), the authors proposed a method for modeling the power amplifier that is based on spars delay taps. Based on this technique, all the memory effects of the power amplifier can be modeled. They use the parameter that indicates what percentage of the memory is modeled. This parameter is called the memory effect modeling ratio (MEMR) and is defined as

$$MEMR_m = 1 - \frac{\left\|E^{(m)}\right\|_2}{\left\|E^{(0)}\right\|_2} \tag{3.45}$$

where from Equation (3.45), E is the error between the measured and simulated data, and can be defined as

$$E^{(m)} = Y - Y^{(m)} \tag{3.46}$$

In this case, the estimated rms error considering N consecutive time data points for the memory polynomial model with $m + 1$ branch can be acquired as

$$rmse[m] = \left(\frac{1}{N}\sum_{k=0}^{N-1}\left|e^{(m)}\left[l - k\right]\right|^2\right)^{1/2} \tag{3.47}$$

The value of MEMR is 0 when the power amplifier is without memory effects and is 1 when the power amplifier includes memory. Here we use the power amplifier model, which was introduced and applied by Ku and Kenney (2003) for comparison with the designed power amplifier, and shows the amount of reduction in ACLR for different MEMR values.

There is another large class of power amplifier models (Saleh, 1981) that is based on frequency-dependent AM-AM and AM-PM conversions of the power amplifier. However, they are usually obtained from single-tone measurements and are difficult to extract from practical baseband inputs and outputs. Therefore, they are not considered here.

3.6 Power Amplifier Simulations

Microwave Office design software tool is used for the design of power amplifiers. The Microwave Office environment provides the ability to significantly improve the performance, capacity, and accuracy of designs. In Figure 3.18, the schematic design of a power amplifier is

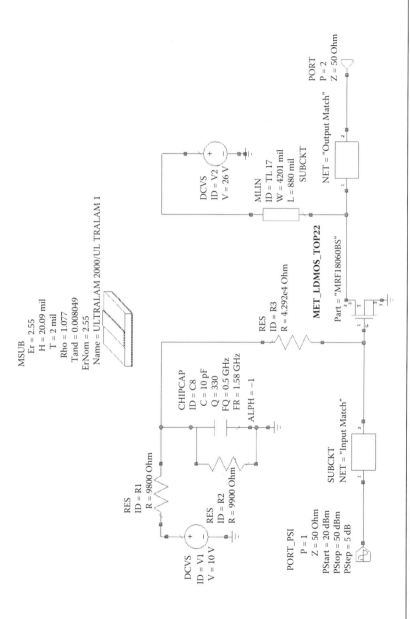

Figure 3.18   Schematic of MRF1806 power amplifier using the Microwave Office software tool.

shown, which is the design of just power amplifier, and later the results will be used in MATLAB® with the predistortion. It is Motorola's LDMOS that is operated at 1.93 GHz to 1.99 GHz with operation up to output of 60 W. The biasing is used from the application circuit in the datasheet, as here the concentration is more on the matching circuit and design of the power amplifier to work with a 60 MHz bandwidth. The biasing of this amplifier is according to the datasheet, as the focus here is not to design the biasing circuit. It should be noted that in this chapter and Chapter 6, the results are based on this power amplifier and in Chapter 4 the comparison and results are based on the actual power amplifier from Mini-Circuits. The input and output matching are accurately designed. The results of input and output matching are shown in Figure 3.19. The power amplifier is designed to cover the 60 MHz bandwidth from 1.93 GHz to 1.99 GHz. The gain parameter S21, which is 13.5 dB, should be flat in all the bandwidth. It is always a trade-off between the gain parameter S21 and input and output matching that represent with S11 and S22. Having the flat gain in all the bandwidth of one of the parameters, S11 or S22, should be sacrificed. As shown in Figure 3.20, the S22 is around −4 dB to −10 dB and is higher than S11, which is at less than −20 dB range.

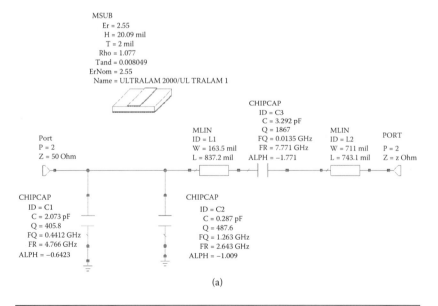

(a)

Figure 3.19   MRF1806 power amplifier matching: (a) input matching.

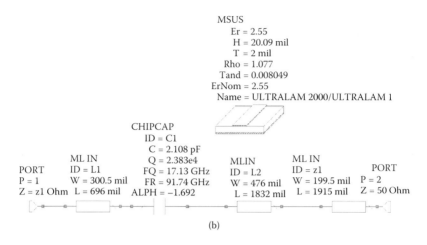

(b)

Figure 3.19 (*Continued*)   MRF1806 power amplifier matching: (b) output matching.

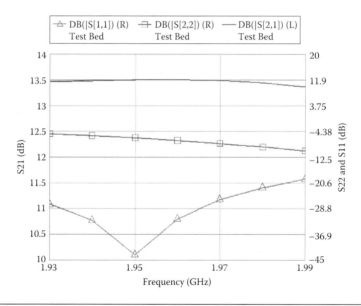

Figure 3.20   S-parameter curves of an MRF1806 power amplifier.

The MRF1806 power amplifier has quite a wide linear region, which is shown in the AM-AM characteristic curve in Figure 3.21a.

It is obvious that the gain of the power amplifier has changed in the saturation area and the power amplifier behaves nonlinearly. The AM-PM characteristic of a simulated power amplifier is shown in Figure 3.21b, and the nonlinear behavior is presented by a variation of the output phase and by increasing the input power.

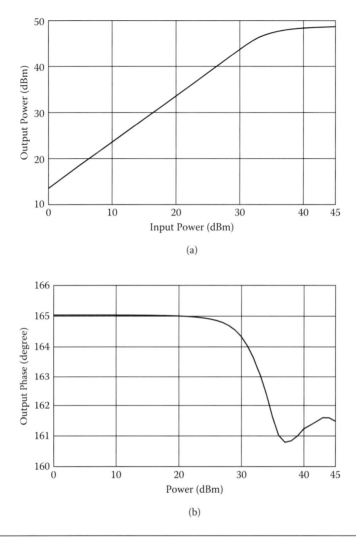

Figure 3.21 The characteristics of an MRF1806 power amplifier: (a) AM-AM, (b) AM-PM.

After the matching results are confirmed, then the AM-AM and AM-PM characteristics of this power amplifier are generated and the 4096 input and output data samples are imported to MATLAB for further analysis. These samples are used to model the power amplifier with memory. The flowchart of the power amplifier with memory is shown in Figure 3.22. Figure 3.23 shows the AM-AM and AM-PM characteristics of this power amplifier that are based on the memory polynomial model, which was discussed in Chapter 1. These characteristics are extracted with a two-carrier

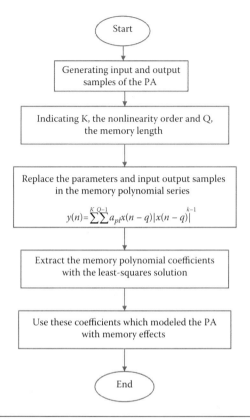

Figure 3.22   Flowchart of a power amplifier model with dynamic memory effects.

WCDMA signal with 10 MHz carrier spacing that is generated from Microwave Office. It shows the scattering of samples because of the memory effects. When there are more memory effects, these samples will be scattered more, then the digital predistortion technique should have more iterations of the adaptation algorithm to compensate for such effects. It is clear in Figure 3.23a that the AM-AM characteristic is not linear when the input amplitude is increased. And also in Figure 3.23b the curve bends. This is because of the nonlinear characteristics of the power amplifier, which here it is assumed that the order of nonlinearity is $K = 3$ and the memory length is $Q = 2$. It should be noted that all the input and output samples in the simulations are normalized. The x and y axis are normalized in these figures. These values are real values and after normalizing these values to their maximum values, the results are shown in Figure 3.23, which is scaled.

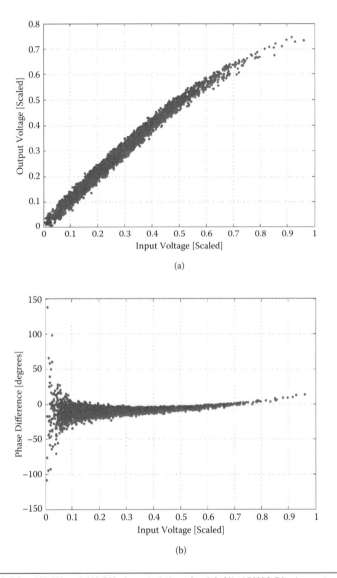

Figure 3.23 AM-AM and AM-PM characteristics of a 1.9 GHz LDMOS PA when a two-carrier WCDMA signal is applied and when $K = 3$ and $Q = 2$: (a) input voltage versus output voltage, (b) input voltage versus phase difference.

3.7 Summary

In this chapter, power amplifiers in wireless communications systems were studied. First, power amplifier efficiency is defined and power amplifier behavior and its characteristics such as AM-AM and AM-PM are discussed. Different power amplifier classes are analyzed and the optimum power amplifier to be used in wireless communications systems is investigated. An example of a designed power amplifier and its characteristics was presented.

References

Benedetto, S., Biglieri, E., and Daffara, R. 1979. Modeling and performance evaluation of nonlinear satellite links, a Volterra series approach. *IEEE Transactions on Aerospace and Electronic Systems* 15:494–507.

Benedetto, S., and Biglieri, E. 1983. Nonlinear equalization of digital satellite channels. *IEEE J Select Areas Commun* 57–62.

Cripps, S. C. 2006. *RF Power Amplifiers for Wireless Communications.* Norwood, MA: Artech House.

Gupta, R., Ballweber, B. M., and Allstot, D. J. 2001. Design and optimization of CMOS RF power amplifiers. *IEEE Journal of Solid-State Circuits* 36(2):166–175.

Heutmaker, M. S. 1997. The error vector and power amplifier distortion, in Proceedings of IEEE Wireless Communications Conference, pp. 100–104.

Kenington, P. B. 2000. *High Linearity RF Amplifier Design.* Norwood, MA: Artech House.

Kim, J., and Konstantinou, K. 2001. Digital predistortion of wideband signals based on power amplifier model with memory. *Electronic Letters* 37:1417–1418.

Krauss, H. L., Bostian, C. W., and Raab, F. H. 1980. *Solid State Radio Engineering.* New York: Wiley.

Ku, H., and Kenney, J. 2003. Behavioral modeling of nonlinear RF power amplifiers considering memory effects. *IEEE Transactions on Microwave Theory and Techniques* 51(12).

O'Droma, M. S., Portilla, J., Bertran, E., Donati, T. J., Brazil, S., Rupp, M., and Quay, R. 2004. Linearization issues in microwave amplifiers. Proceedings of European Gallium Arsenide and Other Compound Semiconductors Application Symposium (GAAS'04–EUMW).

Patel, J. 2004. Adaptive digital predistortion linearizer for power amplifiers in a military UHF satellite. Department of Electrical Engineering, University of South Florida, March.

Pinal, P. L. G. 2007. Multi look-up table digital predistortion for RF power amplifier linearization. Department of Signal Theory and Communications, Universitat Politecnica de Catalunya, Barcelona, Spain.

Raab, F. H., Asbeck, P., Cripps, S., Kenington, P. B., Popovic, Z. B., Pothecary, N., Sevic, J. F., and Sokal, N. O. 2002. Power amplifiers and transmitters for RF and microwave. *IEEE Transactions on Microwave Theory and Techniques* 50(3):814–826.

Raich, R., and Zhou, G. T. 2002. On the modeling of memory nonlinear effects of power amplifiers for communication applications, in Proceedings of the IEEE Digital Signal Processing Workshop, October.

Saleh, A. A. M. 1981. Frequency-independent and frequency-dependent nonlinear models of TWT amplifiers. *IEEE Transactions on Communications* 29:1715–1720, November.

Vuolevi, J. 2003. *Distortion in RF Power Amplifiers.* Norwood, MA: Artech House.

Varahram, P., Mohammady, S., Hamidon, M. N., Sidek, R. M., and Khatun, S. 2009. Digital predistortion technique for compensating memory effects of power amplifiers in wideband applications. *Journal of Electrical Engineering* 60(3):129–135.

Voelker, K. 1995. Apply error vector measurements in communications design. *Microwave and RF* 143–152.

Vuolevi, J. 2003. *Distortion in RF Power Amplifiers.* Norwood, MA: Artech House.

Vuolevi, J. H. K., Rahkonen, T., and Manninen, J. P. A. 2001. Measurement technique for characterizing memory effects in RF power amplifiers. *IEEE Transactions on Microwave Theory and Techniques* 49(8):1383–1389.

Wright, A., and Nesper, O. 2000. Multi-carrier WCDMA base-station design considerations for amplifier linearization and crest factor control. Technology White Paper, PMC-Sierra, Santa Clara, CA.

4

PEAK-TO-AVERAGE POWER RATIO

4.1 Introduction

As explained in Chapter 2, the orthogonal frequency division multiplexing (OFDM) signal is created by the sum of several sinusoidal signals. This process is performed by the inverse fast Fourier transform (IFFT) in the OFDM transmitter.

The basic equations of the FFT and the IFFT can be given by

$$X(k) = \sum_{n=1}^{N-1} x(n) e^{-j2\pi kn/N}, \ k = 0, ..., N-1 \tag{4.1}$$

$$x(n) = \frac{1}{N} \sum_{n=1}^{N-1} X(k) e^{-j2\pi kn/N}, \ n = 0, ..., N-1 \tag{4.2}$$

where N is the transform size or the number of sample points in the data frame and $j = \sqrt{-1}$. $X(k)$ is the frequency output of the FFT at kth point where $k = 0, 1, ..., N-1$, and $x(n)$ is the time sample at nth point with $n = 0, 1, ..., N-1$.

Due to the symmetric of the exponential matrix $e^{-\frac{j2\pi nk}{N}}$, it can be represented as a twiddle factor that is shown with W_N^{nk}. The computation can be performed faster by using a twiddle factor as it depends on the number of points used and there is no need to recalculate it and the values can be referred to a matrix of twiddle factors. As the transform time is very crucial in the FFT process, there is always a trade-off between the core size and the transform time.

Here, an example is presented to give an explanation of how the process of IFFT in an OFDM transmitter leads to high peak-to-average power ratio (PAPR) in OFDM signals. As shown in the block diagram in Figure 4.1, the first random data between 0 and 3 with

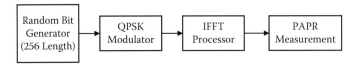

Figure 4.1 The block diagram of an OFDM example.

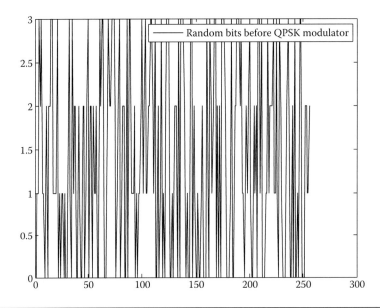

Figure 4.2 The random bits before modulation.

a length of 256 is generated. This random data before introducing to the modulator block is illustrated in Figure 4.2.

As shown in Figure 4.1, the next block after the random bit generator is the modulator block, which modulates the random data and passes the signal to the IFFT processor. Here quadrature phase-shift keying (QPSK) modulation is applied. The modulated signal with a length of 256 can be seen in Figure 4.3a. It should be noted that the inverse process, which is demodulation, is required in the receiver side to facilitate extracting the original data.

In order to create an OFDM symbol, it is necessary to use an IFFT processor. As shown in the block diagram of Figure 4.1, after the QPSK modulator, the signal is introduced to an IFFT processor. The IFFT formula presented in Equation (4.2) is executed on the input signal and the output signal is shown in Figure 4.3b. As can be seen, there are many high peaks in this typical signal. To explain the

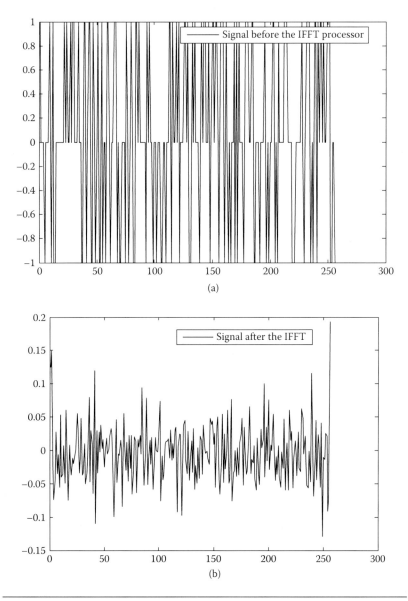

Figure 4.3   The random bits: (a) after QPSK modulation, (b) after the IFFT processor.

issue of high PAPR in this signal, the PAPR is measured. The first thing to do to calculate the PAPR is to measure the power of the signal, which is equal to the absolute of power 2 of the signal that can be presented as

$$Power\ of\ Signal = \sqrt{Signal^2} = \sqrt{S^2} \qquad (4.3)$$

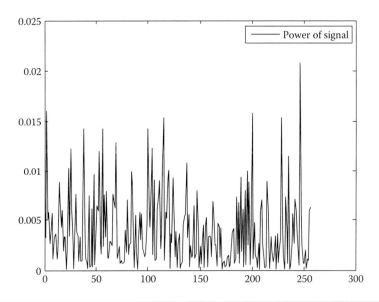

Figure 4.4 The power of the signal.

The power of the signal is shown in Figure 4.4. There are many high peaks, which indicate that the maximum of the signal power is very high. As shown in Figure 4.4, the maximum power can be as high as 0.021 near the end of the signal. It should be noted that in the OFDM transmitter in order to prevent interferences, the last portion of the signal is copied to the beginning of the signal, this process is called cyclic prefix (CP).

Some of the values for the power of the signal are presented in the following equation:

Power of Signal = [..., 0.00584040525900218, 0.00390533839357974, 0.00270909019431493, 0.000479237847018842, 0.0207854700008689, 0.00544550141335431, 0.00244292856689376, 0.000680893271378500, 0.000669839284206245, 0.00200547804530024, 0.000153502560362422, 0.00115804783244301, 0.000989621129863748, 0.00592046875968642, 0.00621618545560074]   (4.4)

It can be observed that the maximum of these values is approximately equal to 0.0208, which is underlined. This can also be seen in Figure 4.4 that in the 246th subcarrier, a very high peak appears.

The formula for calculating the PAPR can be presented as

$$PAPR = \frac{Maximum\ of\ the\ Power\ of\ Signal}{Mean\ of\ the\ Power\ of\ Signal} \qquad (4.5)$$

Generally, the PAPR is defined as a logarithmic ratio of maximum or peak power to average power, $E\{.\}$, as follows:

$$PAPR_{dB} = 10\log\left(\frac{Max\{S(t)\}}{E\{S(t)\}}\right) \qquad (4.6)$$

In communication systems, the logarithmic scale helps to cover a large range of values and the PAPR in logarithm format, denoted as dB, is more manageable to plot.

Following Equations (4.5) and (4.6), the next step is to evaluate the mean of power of the signal, $E\{.\}$. The mean of power can be extracted by using the following formula:

$$E\{s(t)\} = Mean\ of\ Power = M_p(s_1, s_2, s_3, ..., s_n) = \left(\frac{1}{n}\sum_{i=1}^{n} S_i^P\right)^{\frac{1}{p}} \qquad (4.7)$$

where p is a real number and n is the length of the signal. In the current example, the length of the signal is $n = 256$. From Equations (4.4) and (4.6), it can be observed that the mean of power is equal to 0.0039.

To calculate an accurate PAPR value for the examples signal, 10^5 symbols of OFDM are captured and analyzed. As a result, 10^5 numbers of PAPR values are computed. This behavior can be illustrated using a complementary cumulative distribution function (CCDF) graph. The CCDF is a probability function that is used to study the PAPR performance of the OFDM signal. Generally, the CCDF denotes the probability that the PAPR of a data symbol exceeds a predefined threshold as expressed by (Han and Lee, 2005; Heo et al., 2009):

$$probability\ (PAPR \rangle z) = 1 - probability\ (PAPR \leq z)$$

$$= 1 - F(z)^N = 1 - (1 - \exp(-z))^N \qquad (4.8)$$

$$F(z) = 1 - \exp(z)$$

where N is the number of subcarriers and z is the threshold. Basically, this probability function is used as a graph to determine the ability of an algorithm in reducing the PAPR of the OFDM signal and the PAPR is usually compared to an unmodified OFDM signal at 0.01% CCDF, which is shown by the 10^{-4} CCDF in the horizontal vector of graphs. A typical OFDM signal without any PAPR reduction technique has about 8 dB to 13 dB PAPR at 10^{-4} CCDF (Raab et al., 2002). Therefore, when a PAPR reduction technique is applied to the OFDM system, it is expected to reduce the 13 dB PAPR to some lower value. According to IEEE STD 802.16–2005, the reduction should be at least 3 dB.

To illustrate the CCDF of the example signal, the semilogarithmic function is used. Generally, a semilog graph or semilog plot is a way of visualizing data that are related according to an exponential relationship. Normally, one axis is plotted based on a logarithmic scaled parameter and the other axis presents the probability. The reason for using a semilog plot is that the large range of values can be analyzed and compared. The semilog helps to bring out features in the data that would not be easy to do with a basic logarithmic plot. The CCDF graph of the example signal is shown in Figure 4.5.

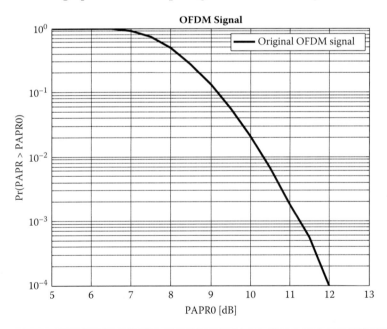

Figure 4.5 The CCDF graph of an example OFM signal.

As shown in Figure 4.5, the horizontal axis is showing the PAPR value in decibels. The vertical axis is showing the CCDF known as a probability of the PAPR. To understand this figure, the point that the curve crosses the horizontal axis should be considered. According to Figure 4.5, this curve crosses on 12 dB. This means that the probability of having higher than 12 dB PAPR in this signal is about 10^{-4}, which is equal to 0.01%. This value is considered as a high PAPR since it will cause dilemma for the power amplifier (PA), which is a power limited component. This phenomenon is explained in the next section.

4.2 The Effect of High PAPR on Power Amplifiers

As explained in Chapter 2, due to the constructive interference between sinusoids, as shown in Figure 4.6, when all points of the sinusoids achieve the maximum value at the same time, this will cause the output envelope to suddenly rise, which causes a "peak" in the output envelope and as a result very high peaks will be structured.

Figure 4.6 shows a sum of five sinusoid signals of sin(x), sin(x + pi), sin(x − pi), sin(2x), and sin(x/2). As can be observed from this simple example, high peaks will be generated due to the constructive characteristic. In this example, the amplitude changes are not considered.

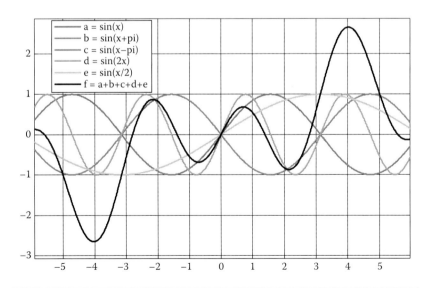

Figure 4.6 The constructive phenomena between four sinusoids.

If the amplitude changes are also considered, as a result of destructive interference, the average of these signals will be very low.

In OFDM signals, the average power of the signal might be as low as zero, hence, due to the presence of a large number of independently modulated subcarriers in an OFDM system, the ratio between peak power and average power will be high (Higashinaka et al., 2009). This parameter can be defined as a relation between the maximum power of a sample in a given OFDM transmit symbol divided by the average power of that OFDM symbol, presented as

$$Peak\ to\ Average\ Power\ Ratio\ (PAPR)\ or\ (PAR) = \frac{Peak\ Power}{Average\ Power}$$

$$= \frac{Max\ of\ Signal}{Average\ of\ Signal}\ Crest\ Factor\ (CF) = \sqrt{PAPR}$$

$$(4.9)$$

Peak-to-average power ratio (PAPR), also referred to as PAR or squared crest factor (CF), is known as a major design challenge in OFDM systems (Bauml et al., 1996; Krongold and Jones, 2003; Wei et al., 2006). The explanation for the problem caused by high PAPR signal to high power amplifiers (HPAs) is that power amplifiers are peak power limited. Here, an example PA is tested and the characteristics are explained. The tested PA is ZVE-8G+ from Mini-Circuits. This PA operates in class AB and so it is suitable for telecommunication application.

A typical test has been performed by using a vector signal analyzer (VSA) and signal generator in order to capture the amplitude characteristic known as AM-AM and phase characteristics known as AM-PM. For this test, 2048 samples were passed through the PA to analyze the behavior of the PA. The AM-AM and AM-PM characteristics of PAs are presented in Figure 4.7a,b, respectively.

According to Figure 4.7a, the PA characteristic can be categorized into three parts. As can be seen in the AM-AM plot, when the input signal is less than 0.4 in normalized unit of voltage, the PA shows a linear behavior. However, if the input signal exceeds this point, the PA characteristic follows the curve-shape path, which indicates a very nonlinear behavior. When this example PA is operated below 0.4, by any increase in input signal, the output signal is increased linearly

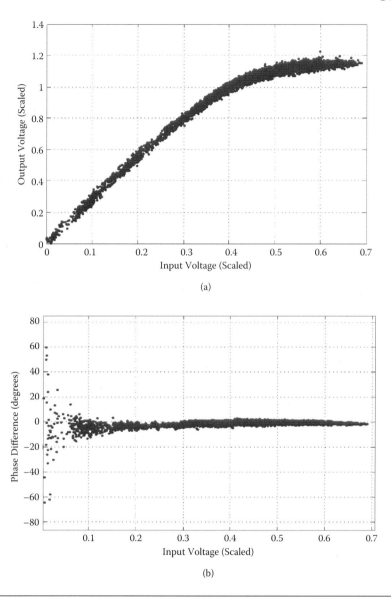

Figure 4.7   The characteristics of the ZVE-8G PA: (a) AM-AM, (b) AM-PM.

and correspondingly. However, if the PA is forced to operate between 0.4 and 0.6, as observed from the characteristic, the output signal will not be increasing accordingly and the amplification is very minor. This leads to less efficiency in the PA and transmission. The other main drawback of this phenomenon is that when the PA shows nonlinear behavior, the output signal will have distortions very close to the main spectrum, which is very difficult to compensate and will interfere with adjacent channels. This out-of-band distortion or spreading the spectrum can be measured by the adjacent channel power ratio (ACPR) metric. There is another impact from the nonlinearity of PAs, which is known as in-band distortion, which can be measured by the error vector magnitude (EVM) metric.

When a single carrier signal such as frequency division multiplexing (FDM) is introduced to this PA, the PA nonlinearity can be compensated with many existing linearization techniques such as feedforward or predistortion and therefore the efficiency of the PA can be protected. However, if a multicarrier signal with high PAPR, such as an OFDM signal, is introduced to this PA, because of a very wide dynamic requirement of this signal, the PA is forced to operate in a nonlinear region. This response is shown in Figure 4.8.

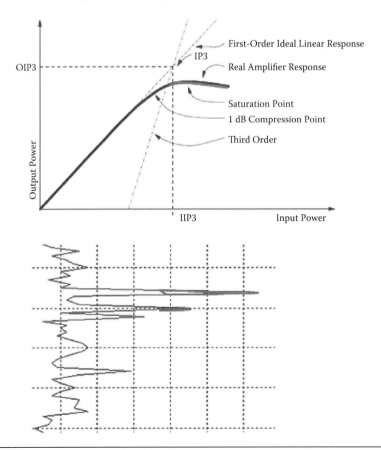

Figure 4.8   The nonlinear response of a PA to an input signal with high PAPR.

There are some PAs with a wide dynamic linear region (class AB), but they are generally expensive, consume more power, and are less efficient (Cooper, 2008; Sharma and Basu, 2010; Varahram et al., 2010).

Hence, in order to have a high-efficiency OFDM signal and extended battery life especially in mobile systems, the PAPR of OFDM signal must be reduced and the linearity of the PA should be maximized. Many PAPR reduction algorithms have been developed and are discussed next.

4.3 PAPR Reduction Techniques

Several techniques have been developed to reduce PAPR of the OFDM signal. There are two main categories for these techniques: distortion-based methods (which means that applying these methods result in out-of-band distortion) and distortionless methods (there is no out-of-band distortion). The first category includes clipping (May and Rohling, 1998), windowing (Van Nee and De Wild, 1998), envelope scaling (Foomooljareon et al., 2003), random phase updating (Nikookar and Lidsheim, 2002), peak reduction carrier (Tan and Wassell, 2003), companding (Cao et al., 2007; Chang et al, 2010; Hao and Liaw, 2006, 2008; Kim et al., 2008), and other modified versions of these methods.

Clipping is a simple technique for PAPR reduction, where in the transmitter, the signal is clipped to a desired level and the phase information remains unchanged. In a windowing technique, a large signal peak is multiplied with a certain frame. The envelope scaling method is an algorithm to reduce PAPR by scaling the input envelope for some subcarriers before they are sent to IFFT. In the random phase updating algorithm, some random phases are generated and assigned for each carrier. The process of updating is continued until the peak value of the OFDM signal is below the threshold. The peak reduction carrier involves the use of a higher order modulation scheme to represent a lower order modulation symbol (Vijayarangan and Sukanesh, 2009). The companding technique is used to compress and expand the OFDM signal in order to reduce PAPR. Speech processing is the main application of the companding method because it has less frequent peak problems.

The second category of PAPR reduction methods is called the distortionless techniques. These methods have significant PAPR performance without causing nonlinear distortion. However, they typically incur large computational complexities and sometimes side information transmission. Moreover, these methods usually require receiver-side modifications that may be incompatible to existing communication systems. Such approaches include coding (Jones et al., 1994; Kwon, Park, et al., 2009), partial transmit sequence (PTS) (Chen, 2010; Gao et al., 2009; Kang et al., 1999; Muller and Huber, 1997c), selected mapping (SLM) (Bauml et al., 1996), dummy signal insertion (DSI) (Qian et al., 2005; Ryu et al., 2004), tone injection and tone reservation (Tellado, 2000a), interleaving (Jayalath and Tellambura, 2000), and active constellation extension (ACE) (Krongold and Jones, 2003).

Most of the recent research concentrates on modified SLM and PTS methods (Ghassemi and Gulliver, 2010; Hong and Har, 2010; Jeon et al., 2011; Kim et al., 2006; Naeiny and Marvasti, 2011; Wang and Liu, 2011). As discussed in this chapter, most of the modified methods reduce PAPR at the expense of complexity in the transmitter or degrading the spectrum efficiency of the system. Hence, there is good scope to design a new method to overcome previous drawbacks and enhance the PAPR performance.

In addition to the efficiency and the complexity of the OFDM system, the PAPR reduction method must also meet the PA nonlinearity issue. The basic way to avoid operating the PA in the nonlinear region is to back off the input signal (reduce the input power) by an amount proportional to the PAPR, which leads to a decrease in output power and as a result the power efficiency will be degraded (Gilabert et al., 2009; Muller and Huber, 1997c; Pratt et al., 2006; Qian et al., 2005). The other outcome of PA nonlinearity is the distortion that appears around the signal spectrum, which leads to interference with adjacent channels, which is distinctly unwanted in communication especially in green technology.

The solution to have high-efficiency amplification and enhance the linearity of the PA is to use the linearization technique. Several linearization techniques have been proposed, such as feedforward (Youngoo and Bumman, 1999), Cartesian feedback (Faulkner, 2000), and digital predistortion (DPD) (Chang and Yang, 2009; Faulkner et al., 1994; Gilabert et al., 2009; Varahram et al., 2010). Among all the linearization methods, the baseband DPD technique is very popular because it can be applied to many kinds of power amplifiers and different kinds of modulated signals (Varahram et al., 2010). Another advantage of DPD is the cost, since it is performed in the digital domain. The cost and simplicity makes it a very flexible algorithm.

As discussed earlier in this chapter, PAPR reduction algorithms are generally divided into two groups of distortion based and distortion-less based (Han and Lee, 2005). These are discussed in detail next.

4.3.1 Distortion-Based PAPR Reduction Techniques

Clipping the OFDM signal before amplification is the simplest method to limit PAPR. Deliberate clipping (Ochiai and Imai, 2002) may cause large out-of-band and in-band distortion, which degrades the system performance. Other methods in the distortion-based group are repeated clipping and filtering (Leung et al., 2002), peak power suppression, weighted multicarrier transmission, companding (Chang et al., 2010; Huang et al., 2004; Jiang et al., 2005), and other simple PAPR reduction algorithms with distortion-like peak windowing and peak cancellation. These methods reduce the PAPR effectively; however, they introduce distortion. Therefore, the system performance should be carefully controlled, especially while it is working with a nonlinear PA. These techniques directly reduce high peaks by distorting the signal prior to amplification and generally require less computation. But still there is a trade-off among PAPR reduction capability in terms of spectral regrowth and bit error rate (BER). Next, we briefly explain some of these methods.

4.3.1.1 Clipping Method
Clipping was one of the first techniques proposed for PAPR reduction, where the signal at the transmitter is clipped to a desired level without modifying the phase information. The amplitude limitation of the time domain signal affects the received signal on all subcarriers (in-band distortion) and creates out-of-band emission (Gregorio, 2007; Sulaiman et al., 2007). In 1998, May and Rholing proposed this method. After them, many modifications have been developed, including by Gurung et al. (2008), Urban and Marsalek (2008), Xue et al. (2009), Kwon, Kim, et al. (2009), and Deng and Zhong (2008). They all investigated the relation between PAPR

and the clipping noise level and the performance of some popular PAPR reduction schemes, and some of them have shown that the dependence between the clipping noise and PAPR is weaker when the clipping threshold is decreased. However, the impact of clipping is significant only when the threshold is low (Sulaiman et al., 2007). The following equation shows the function of amplitude clipping, which is known as a soft limiter (Han and Lee, 2005; Ochiai and Imai, 2002)

$$g(x) = \begin{cases} x, |x| < CL \\ CLe^{j\angle x}, |x| \geq CL \end{cases} \tag{4.10}$$

where CL is the amplitude threshold and $g(x)$ is the output signal. As shown in Figure 4.9, the peak of the signal will be cut off if the signal passes the threshold value and the output will be equal to x, which is the input signal.

Armstrong (2002) proposed a method based on the repeated process of clipping and filtering. According to the block diagram in Figure 4.10, this technique adds extra FFT and IFFT blocks to the system.

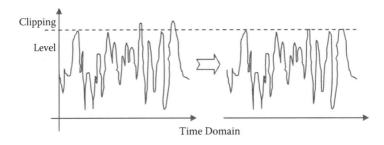

Figure 4.9 A clipped OFDM signal in time domain.

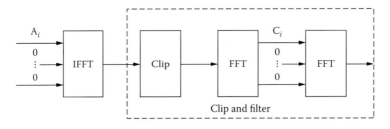

Figure 4.10 The clipping and filtering method for PAPR reduction. (From Armstrong, J. 2002. *Electronics Letters* 38(5):246–247. With permission.)

This design can reduce the PAPR by about 2 dB when only one time clipping and filtering is used. When repeating in the algorithm is increased, the PAPR reduction is improved. However, this repetition increases the process time and additional IFFT and FFT increase the computational complexity of the transmitter. Armstrong (2002) claimed that by applying the filtering process to clipping, out-of-band distortion is eliminated but in a practical system in-band distortion exists in the form of the clipping noise.

4.3.1.2 Windowing Method In the windowing technique, the Gaussian-shaped window is multiplied with the peak signal in order to reduce it. The resulting spectrum is a complication of the original OFDM spectrum with the spectrum of the applied window. The window should not be too long in the time domain due to BER degradation.

In 1998, Van Nee and De Wild (1998) suggested that since large PAPRs occur infrequently, peaks can be removed at the cost of a slight amount of self interference. Luo et al. (2008) claimed that the side lobes can be suppressed by raised cosine (RC) windowing in time domain, while the intercell interference can be suppressed by reserving a guard band between neighboring frequency channels. Higher roll-off factor (ROF) of the RC window leads to better side lobe suppression and allows a smaller guard band size, increasing the spectral efficiency. The windowing method can reduce PAPR by about 4 dB and the loss of signal-to-noise ratio (SNR) is limited to about 0.3 dB.

4.3.1.3 Companding Method Companding transform (CT) reduces the PAPR without changing the number of subcarriers, frame format, and constellation type (Aburakhia et al., 2009). This technique is a combination of compressing and expanding, and is simple to be implemented in integrated circuit design (Cao et al., 2007; Hao and Liaw, 2008). Rostamzadeh et al. (2008) proposed a novel adaptive companding transform scheme to effectively reduce the PAPR of OFDM and wavelet packet division multiplexing (WPDM) signals. Wang et al. (1999) proposed a simple technique to compand the signal before the analog stage (Wang and Ku, 2006). Gaussian distribution is used to model the signal and it is shown that high peaks occur infrequently. At the receiver, the received signal is passed through an analog-to-digital converter and expanded. As mentioned before, the companding method is used in speech processing. Due to companding, the quantization error for large signals is considerably large, which degrades the BER performance of the system (Hosseini et al., 2006). In Takyu et al. (2006), each symbol is divided into groups of consecutive segments or clusters. Each cluster is multiplied by a weight equal to the inverse of the largest instantaneous power within the cluster. The receiver must guess the weight of each cluster, which results in BER degradation (Vijayarangan and Sukanesh, 2009).

4.3.2 Distortionless-Based PAPR Reduction Methods

When distortionless-based methods are applied to the system to reduce PAPR, no distortion is added; however, computational complexity and the number of side information bits are usually increased and receiver design matching with the transmitter modifications is required, which should be compatible to existing communication systems. Such approaches include coding (Jones et al., 1994; Kwon, Park, et al., 2009), selected mapping (SLM) (Bauml et al., 1996), dummy signal insertion (DSI) (Ryu et al., 2004), partial transmit sequence (PTS) (Muller and Huber, 1997a; Wu, Wang et al., 2010), interleaving (Jayalath et al., 2000), active constellation extension (ACE) (Krongold and Jones, 2003), tone injection or tone reservation (Tellado, 2000a), and other modified methods (Qian et al., 2005).

These distortionless-based techniques are also known as signal scrambling techniques. Signal scrambling techniques are all variations on how to scramble the codes to control the PAPR (Vijayarangan and Sukanesh, 2009). Coding techniques can be used for signal scrambling. More practical solutions of the signal scrambling techniques are SLM, PTS, and block coding. Jones and Barton proposed a block coding scheme for PAPR reduction in 1994. Other methods using different types of codes were proposed by Ochiai and Imai (1997). SLM and PTS methods are the most promising techniques in terms of PAPR reduction. However, they both have the problem of side information and receiver modification. According to Baxley and Zhou (2007), the SLM technique has less total complexity compared to the PTS method.

4.3.2.1 Coding Method   In the coding method, the data is coded and so the set of permissible code words does not contain those that create

high peaks. The coding method has three steps. The first step is the selection of code words. The second step is the selection of the sets of code words considering criteria of efficient implementation. The third step is the selection of sets of code words for error deduction and correction (Kwon, Park, et al., 2009; Vijayarangan and Sukanesh, 2009).

Wilkinson and Jones proposed a modified block coding scheme for the PAPR reduction in 1995. Huang et al. (2008) proposed a new method to construct nonbinary codes that achieve excellent performance, match well with the modulation, and can be encoded in linear time and in parallel. Coding-based methods have moderate PAPR performance. However, the complexity of the transmitter and receiver is relatively high, which is not suitable for practical applications.

4.3.2.2 Active Constellation Extension   Douglas proposed the active constellation extension (ACE) method (Jones, 1999; Jones et al., 1994) to reduce peaks of the OFDM signal. ACE uses active (data-carrying) channels that dynamically move outer constellation points, within margin-preserving constraints, to minimize the peak magnitude.

The shaded region in Figure 4.11 shows the area that modification on the constellation is safe and it will not degrade the performance of the system. The effect of this modification is to add an adjusted

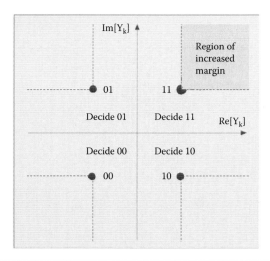

Figure 4.11   The four quadrants bounded by the real and imaginary axes form the decision regions. (From Jones, D. L. 1999. Peak power reduction in OFDM and DMT via active channel modification. Proceedings of 33rd Asilomar Conference on Signals, Systems, and Computers, Pacific Grove, CA, October 24–27. With permission.)

combination of cosinusoids and/or sinusoids in a particular frequency channel to cancel the time domain peaks of the OFDM signal (Jones, 1999). It actually increases the average in Equation (4.9), when peaks are constant, therefore the PAPR will be decreased.

4.3.2.3 Partial Transmit Sequence   The partial transmit sequence (PTS) is a promising technique that works with a probabilistic algorithm for PAPR reduction in OFDM systems (Wu, Wang et al., 2010). Muller and Huber (1997b) first proposed the PTS method as a flexible and effective PAPR reduction method. As shown in the block diagram of ordinary PTS (O-PTS) in Figure 4.12, the data block is divided into nonoverlapping partitions called subblocks, which are rotated with an independent rotation factor. Then, the time domain data with the lowest peak amplitude is generated and transmitted to the receiver. The rotation factor is also transmitted as side information.

By introducing the orthogonal design, a phase factor sequences algorithm can be designed. This algorithm uses an orthogonal phase factor sequences generator (Wang, Wu, et al., 2009). This modified PTS method reduces PAPR by about 4 dB at 10^{-4} CCDF, while 4 IFFTs should be implemented.

A major drawback of PTS is the need of transmitting rotation factors for decoding, which degrade the transmission efficiency (TE) and

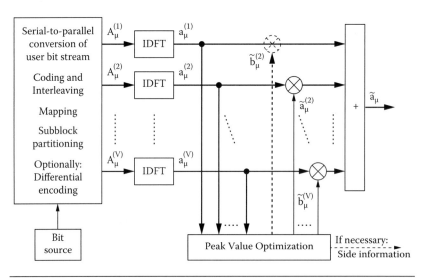

Figure 4.12   Block diagram of PTS in an OFDM system. (From Muller, S. H., and Huber, J. B. 1997. *Electronics Letters* 33(5):368–369. With permission.)

increases the complexity of the system. Giannopoulos and Paliouras (2009) proposed a low-complexity PTS technique for retrieving the weighting factors in the receiver. The modified decoder explores all the permissible combinations of weighting factors in order to identify the factor combination employed by the transmitter, so no additional pilot tones or explicit transmission of side information are required.

Wang, Guo, et al. (2009) presented another modified PTS that is a combination of the linear and nonlinear methods, named PTS, followed by clipping (PTS-clipping). It reduces the PAPR of radio over a fiber and orthogonal frequency division multiplexing (ROF-OFDM) system. The main part of this method is clipping the processed signal whose probability of the peak value has been reduced by PTS technique and then the PAPR value will be reduced. Therefore, the BER will be degraded and the system computational complexity will be increased.

One recent publication by Ghassemi and Gulliver (2010) has claimed to significantly decrease the computational complexity while providing comparable PAPR reduction to O-PTS, even with a small number of stages after PTS partitioning. The autocorrelation function (ACF) was used to develop a modified PTS subblocking technique using error-correcting codes (ECCs). This method can reduce the PAPR by about 3.9 dB at 10^{-4} CCDF. This reduction is slightly degraded compared to the O-PTS method when 4 IFFTs are used for implementation.

Wang et al. (2011) proposed two phase weighting methods with low computational complexity for PTS, named grouping phase weighting (GPW) and recursive phase weighting (RPW), which focus on simplifying the computation for candidate sequences and obtaining the same candidate sequences compared to O-PTS.

The PAPR reduction is similar to the O-PTS method, but it is claimed that the complexity is reduced. In this method, while 8 subblocks are used, which means 8 IFFTs, the PAPR can be reduced by about 5 dB at 10^{-4} CCDF. Increasing the number of IFFTs in OFDM design leads to higher complexity.

4.3.2.4 Enhanced PTS　In order to decrease the complexity of conventional PTS C-PTS, a new phase sequence is generated. The block diagram of the enhanced partial transmit sequence (EPTS) scheme is shown in Figure 4.13.

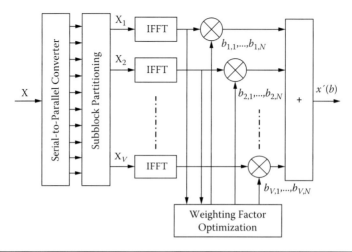

Figure 4.13   Block diagram of the enhanced PTS (EPTS) technique.

This new phase sequence is based on the generation of N random values of $\{1 \; -1 \; j \; -j\}$ if the allowed phase factors is $W = 4$. The phase sequence matrix can be given by

$$
\hat{b} =
\begin{bmatrix}
b_{1,1} & ,\dots, & b_{1,N} \\
\vdots & \vdots & \vdots \\
b_{V,1} & ,\dots, & b_{2,N} \\
b_{V+1,1} & ,\dots, & b_{V+1,N} \\
\vdots & \vdots & \vdots \\
b_{P,1} & ,\dots, & b_{P,N}
\end{bmatrix}_{[P \times N]}
\tag{4.11}
$$

where P is the number of iterations that should be set in accordance with the number of iterations of the C-PTS, and N is the number of samples (IFFT length), and V is the number of subblocks partitioning. The value of P can be calculated as follows:

$$
P = DW^{V-1}, \quad D = 1, 2, \dots, D_N
\tag{4.12}
$$

where D is the coefficient that can be specified based on the PAPR reduction and complexity, and D_N is the value that is specified by the user. The value of P explicitly depends on the number of subblocks V, if the number of allowed phase factors is constant.

There is a trade-off for choosing the value of D: a higher D leads to higher PAPR reduction but at the expense of higher complexity, whereas a lower D gives a smaller PAPR reduction but with less complexity. For example, if $W = 2$ and $V = 4$, then in C-PTS there are eight iterations and hence $P = 8D$. If $D = 2$, then $P = 16$ and both methods have the same number of iterations. But when $D = 1$, then the number of iterations to find the optimum phase factor will be reduced to four and this will result in complexity reduction. The main advantage of this method over C-PTS is the reduction of complexity while at the same time maintaining the same PAPR performance. In the case of C-PTS, each row of the matrix $\hat{b}$ contains the same phase sequence while each column is periodical with period V, but in the proposed method each element of matrix $\hat{b}$ has different random values.

The other formats that the matrix in Equation (4.11) can be expressed are as follows:

$$\hat{b} = \begin{bmatrix} \overbrace{b_{1,1},...,b_{1,N/P}}^{P} & ,..., & b_{1,1},...,b_{1,N/P} \\ \vdots & \vdots & \vdots \\ b_{V,1},...,b_{V,N/P} & ,..., & b_{V,N/P} \\ b_{V+1,1},...,b_{V+1,N/P} & ,..., & b_{V+1,1},...,b_{V+1,N/P} \\ \vdots & \vdots & \vdots \\ b_{P,1},...,b_{P,N/P} & ,..., & b_{P,1},...,b_{P,N/P} \end{bmatrix}_{[P \times N]} \tag{4.13}$$

$$\hat{b} = \begin{bmatrix} \overbrace{b_{1,1},...,b_{1,1}}^{P}, & \overbrace{b_{1,2},...,b_{1,2}}^{P}, & ,..., & \overbrace{b_{1,N/P},...,b_{1,N/P}}^{P} \\ \vdots & \vdots & & \vdots \\ b_{V,1},...,b_{V,1}, & b_{V,2},...,b_{V,2}, & ,..., & b_{V,N/P},...,b_{V,N/P} \\ b_{V+1,1},...,b_{V+1,1} & ,..., & & b_{V+1,N/P},...,b_{V+1,N/P} \\ \vdots & \vdots & & \vdots \\ b_{P,1},...,b_{P,1} & ,..., & & b_{P,N/P},...,b_{P,N/P} \end{bmatrix}_{[P \times N]} \tag{4.14}$$

where Equations (4.13) and (4.14) are the interleaved and adjacent phase sequences matrix, respectively.

As an example, assume $N = 256$, and the number of allowed phase factor and subblock partitioning are $W = 4$ and $V = 4$, respectively. With C-PTS, there are $W^{M-1} = 64$ possible iterations, whereas for the proposed method, in the case of $D = 2$, the phase sequence is a matrix of 128×256 elements according to Equation (4.11). In this case, 64 iterations are required for finding the optimum phase sequence, because each two rows of the matrix in Equation (4.11) multiply pointwise with the time domain input signal x_v with length 2×256. The reduction of subblocks to 2 is because it gives almost the same PAPR reduction as C-PTS with $V = 4$. It should be noted that if $D = 1$ then the complexity increase, and if $D > 2$ then the PAPR reduction is less.

Therefore, the algorithm can be expressed in the following steps:

1. Generate the input data stream and map it to the M-QAM modulation.
2. Construct a matrix of random phase sequence with dimension of $[P \times N]$.
3. Pointwise multiply signal x_v with the new phase sequence.
4. Find the optimum phase sequence after P iterations to minimize the PAPR.

4.3.2.5 Selected Mapping Method The selected mapping (SLM) method reduces the probability of high peaks occurring, but it does not remove the peaks. Bauml et al. (1996) proposed this method where the SLM scheme was introduced as one of the initial probabilistic approaches and compatible with any number of subcarriers. As shown in the block diagram of Figure 4.14, in the SLM method

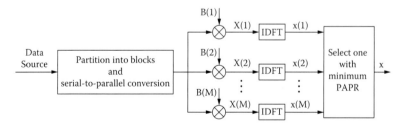

Figure 4.14   The block diagram of the conventional SLM technique. (From Bauml, R. W., et al. 1996. *Electronics Letters* 32(22):2056–2057; Han, S. H., and Lee, J. H. 2005. *IEEE Wireless Communications* 12(2):56–65. With permission.)

the OFDM signal is multiplied by several phase sequences in parallel and the candidate signals are generated. Then the data sequence with the lowest PAPR is selected and will be transmitted (Lee and de Figueiredo, 2010; Ryu et al., 2004).

The main drawback of this technique is transmission of side information; for each data block, it requires the transmission of several side information bits (Tang and Zhang, 2009). The side information is for informing the receiver about the data sequence selection result. This transmission of some extra bits causes some data rate loss. The channel coding method needs to be very powerful to protect these critical side information bits, which results in higher system complexity and transmission delay, and decreases the data rate even further. Le Goff et al. (2008) propose a new SLM method in which no side information needs to be transmitted. Figure 4.14 shows an example of the block diagram of a conventional SLM technique.

Here is an OFDM system with 8 subcarriers and the number of allowed phase sequences is 4. The data block that is going to be transmitted is $X = (1, -1, 1, 1, 1, -1, 1, -1)^T$ and before applying the SLM the PAPR is 6.5 dB. It is assumed that the number of subblocks is $M = 4$. Then the four phase factors are set as follows:

$$\mathbf{B}(1) = (1, 1, 1, 1, 1, 1, 1, 1)^T$$

$$\mathbf{B}(2) = (-1, -1, 1, 1, 1, 1, 1, -1)^T$$

$$\mathbf{B}(3) = (-1, 1, -1, 1, -1, 1, 1, 1)^T$$

$$\mathbf{B}(4) = (1, 1, -1, 1, 1, -1, 1, 1)^T$$

where T indicates the matrix transpose. Between these four modified data blocks $X(m)$, $m = 1, 2, 3, 4$, $X(2)$, has the lowest PAPR of 3.0 dB. So the $X(2)$ is selected and transmitted. In this example, the PAPR of this data block is reduced from 6.5 dB to 3.0 dB. This yields PAPR reduction of 3.5 dB by applying the SLM method. Four IFFT operations and 2 bits of side information are required in this example. As a result, side information of "00" shows that the first sequence is selected and "01" shows that the second sequence is selected. So when the number of sequences is higher, the number of side information bits is more, which is undesirable in terms of design complexity.

For other data blocks, the amount of PAPR reduction may vary, but the reduction of PAPR is possible for all of them (Han and Lee, 2005). When M is higher, the reduction is better but a large number of IFFT blocks are required to generate candidate signals, which leads to a significant increment in the computational complexity.

The procedure of the SLM algorithm can be illustrated by the flowchart in Figure 4.15, which shows that after initialization, random data X is generated and modulation is performed.

After that, the phase sequences are generated and multiplied to the input signal and X_i is created, then the X_i is passed through the IFFT block and x_i denotes the output of IFFT block. Following that the PAPR of every candidate signal x_i is calculated and the sequence with the minimum PAPR is selected. The final step is to transmit the sequence with the lowest PAPR together with side information to the receiver.

As a performance measurement, the complementary cumulative distribution function (CCDF) metric is used to evaluate the PAPR performance. When $M = 2$, only two IFFT blocks are

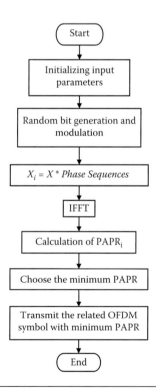

Figure 4.15   Flowchart of the MATLAB program of the SLM method.

required to implement the system and when $M = 16$, the number of required IFFTs are increased to 16 IFFTs.

One of the most used hardware resource consumption units for implementing PAPR techniques is the IFFT in which the IFFT takes around 3850 real additions and multiplications. In the following, the hardware resource consumption has been evaluated by calculating the computational complexity.

A Worldwide Interoperability for Microwave Access (WiMAX) signal with $N = 256$ subcarriers is simulated and the SLM method is applied for PAPR reduction. When the number of carriers is $N = 2^n$, the numbers of complex multiplication n_{mul} and complex addition n_{add} of the conventional SLM (C-SLM) scheme are given by $n_{add} = 2^{n-1}nM$ and $n_{add} = 2^nnM$, respectively, where M is the number of subblocks (Lim et al., 2005). The details of the computational complexity have been studied (Lim et al., 2005; Yang et al., 2008). In Yang et al. (2008), the computational complexity reduction ratio (CCRR) is defined by

$$CCRR = \left(1 - \frac{Complexity\ of\ the\ New\ method}{Complexity\ of\ the\ Conv.\ method} \right) \times 100\% \quad (4.15)$$

If the new method performs with the same complexity as the conventional SLM technique, the CCRR will be 0%. If the new method has less complexity, the CCRR will have the higher value.

Another consideration in applying the SLM method is about recovering the original data at the receiver side. To indicate which data sequence is selected and transmitted, a sequence index called side information also must be transmitted to allow the receiver to generate the original signal. In practice, a very powerful protection is necessary for side information, which will degrade the BER performance. Usually powerful error correction is used to protect side information bits. When $M = 2$, the side information is only one bit. If the bit is "0," it shows that the first signal candidate had the lowest PAPR and so transmitted. If the side information bit is "1," it indicates that the second sequence has the minimum PAPR and is transmitted. When $M = 4$, the number of side information bits becomes 2. The number of required side information bits to send to the receiver can be derived by

$$\text{Number of required side information} = \log_2 M$$

These bits are important to the error performance of the system and they need to be protected by a great error correction. In practice, this increases the system complexity and transmission delay and data rate loss. Le Goff et al. (2008) proposed a modified SLM method that does not require side information bits. It has been shown that the modified method performs very well both in terms of PAPR reduction and BER (Le Goff et al., 2008), however the complexity of the system is noticeably high.

Many modifications have been done by using the SLM method but most of them increase the complexity of the system or degrade the bandwidth efficiency (Chang and Yang, 2009; Irukulapati et al., 2009; Jeon et al., 2011; Kim et al., 2006; Naeiny and Marvasti, 2011; Park et al., 2011; Tao and Li, 2010; Wang et al., 2011; Yang et al., 2009; Zhou and Jiang, 2009).

Tao and Li (2010) combine the C-SLM method with the clipping technique. PAPR reduction is noticeable but the complexity of the system due to the number of IFFTs and also the number of complex multiplications for the phase sequence is the same as the C-SLM method.

Jeon et al. (2011) propose a low-complexity SLM scheme that generates alternative signal sequences by adding mapping signal sequences to an OFDM signal sequence. When the number of subcarriers is 512 and the number of candidate signals is 16, 28, and 40, the PAPR can be reduced to 8.7 dB, 8.5 dB, and 8.3 dB at 10^{-4} CCDF, respectively. It has been claimed to reduce the complexity of the mapping process compared to similar methods while the number of IFFTs is similar to the C-SLM method and the PAPR reduction is not noticeable.

Naeiny and Marvasti (2011) introduce an SLM technique for the PAPR reduction of space-frequency-block-coded OFDM systems with the Alamouti coding scheme. In this method, when the number of candidate signals is 16 and the number of subcarriers is 128, the coded SLM method reduces PAPR by about 3.8 dB at 10^{-4} CCDF, which is about the same result as the C-SLM method. The side information problem is solved by using this coding process but due to 16 IFFTs in the design and decoding process, the complexity of the transmitter and receiver are high.

The idea of coding candidate signals to prevent transmitting side information is also considered in the proposed SLM method (Park et al., 2011).

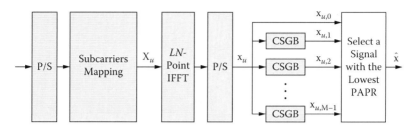

Figure 4.16 The block diagram of a modified SLM. (From Wang, L., and Liu, J., 2011, *IEEE Transactions on Broadcasting* 57(2):299–306. With permission.)

Wang et al. (2011) proposed a modified SLM architecture for improving the PAPR reduction performance and reducing the complexity. It can be seen in Figure 4.16 that following the modulation and IFFT operations, the time-domain signal is processed by M-1 candidate signal generating blocks (CSGBs) to obtain a total of M candidate signals, of which one signal is the original time-domain transmitted signal.

CSGBs generate candidate signals with a similar process as the C-SLM method and hence Log_2M bits side information are required for recovery at the receiver. According to Wang et al. (2011), when one IFFT is used ($M = 1$), the PAPR is reduced from about 11 dB to about 8.4 dB at 10^{-4} CCDF. In other words, at 10^{-4} CCDF, an approximate 2.6 dB reduction is achieved while y = the number of candidate signals M is 16, which means that 16 complex multiplications are required for implementation of this algorithm. In this approach, the number of IFFTs is reduced, which improves the complexity but according to the IEEE standard, the minimum PAPR reduction of 3 dB to 4 dB is not met, which degrade the TE due to power nonlinearity.

In Kim et al. (2006), the PAPR reduction method was designed based on the SLM method and by applying dummy sequences after the IFFT of each candidate signal.

The block diagram for this method, which is illustrated in Figure 4.17, shows that the IFFT is repeated for the dummy sequence. This means that because of the extra IFFTs, the complexity of this system is doubled compared to the C-SLM method.

As shown in Figure 4.17, the Walsh Hadamard transform (WHT) coding scheme is also used to reduce PAPR before the IFFT block. By applying this method, an additional frequency diversity benefit is expected but again the complexity of the transmitter is increased.

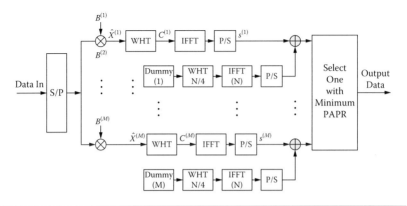

Figure 4.17   The block diagram of the SLM technique with a dummy. (From Kim, S. W., Chung, J. K., and Ryu, H. G. 2006. PAPR reduction of the OFDM signal by the SLM-based WHT and DSI method. IEEE Region 10 Conference (TENCON-2006), Hong Kong, China, November 14–17. With permission.)

Moreover, the receiver requires a special decoding strategy to recover the original signal, which increases the complexity of the receiver too.

In Kim et al. (2006), when the modified SLM method is applied to the OFDM system, the PAPR can be reduced by about 0.5 dB to 1 dB more than the C-SLM and dummy signal insertion (DSI) methods.

4.3.2.6 Tone Reservation Method   The tone reservation (TR) method reserves a small fraction of tones and as a result the PAPR will be reduced. The PAPR reduction performance depends directly on the number of reserved tones, location, and the frequency vector. There is also a trade-off between the complexity of the system and PAPR reduction. When the number of tones is small, the transmitter design is simple but the PAPR performance is not significant and also the data rate is degraded. However, the increase in number of tones will increase the complexity of the system. According to the literature, the TR method is moderate in terms of complexity and has the main advantage of not requiring side information bits (Tellado, 2000a).

4.3.2.7 Dummy Signal Insertion Method   The dummy signal insertion (DSI) method is more efficient than the block coding method and simpler than the PTS and SLM methods. In DSI, some dummy signals are added into the main data signal before the IFFT stage. A block diagram of basic DSI technique is presented in Figure 4.18.

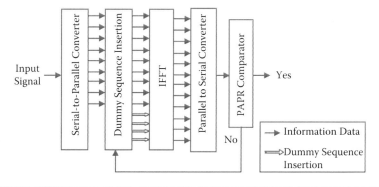

Figure 4.18 Block diagram of the DSI method. (From Ryu, H. G., Lee, J. E., and Park, J. S. 2004. *IEEE Transactions on Consumer Electronics* 50(1):89–94. With permission.)

X IFFT Input										
Divided Data I								**Dummy Bits**		
1	...	$\dfrac{L}{M}$	$\dfrac{L}{M}+1$	...	$\dfrac{2L}{M}$	...	$\dfrac{(M-1)L}{M}+1$	...	L	1	...	M
$I^{(1)}$		$I^{(2)}$			...	$I^{(M)}$				Dummy Bit		

Figure 4.19 Correlation sequence of DSI. (From Ryu, H. G., Lee, J. E., and Park, J. S. 2004. *IEEE Transactions on Consumer Electronics* 50(1):89–94. With permission.)

As shown in Figure 4.18, the serial-to-parallel converter divides the input data D into parallel L bits in the serial. Then, L bits dummy sequence is added. PAPR is compared with a PAPR threshold and if it is lower than the threshold, then it is transmitted. Otherwise, another dummy sequence is inserted into the input data to reduce the PAPR and following the IFFT block the comparator checks the PAPR once more. Several iterations are required to obtain the low PAPR, which takes quite a long time (Ryu et al., 2004). A good example of the DSI method is now provided. If the input signal with length L is $\{I_1, I_2, \ldots, I_L\}$ and the dummy sequence length is M, then the DSI data block using the complementary sequence and the correlation sequence will be as shown in Figure 4.19.

According to Figure 4.19, the difference between the complementary sequence DSI and correlation sequence DSI is in the method of multiplexing the dummy sequence with the input signal. In both conditions, it is not necessary to transmit side information when the dummy sequence is used. At the receiver, the dummy sequence can be discarded after the FFT block, unlike the conventional PTS and SLM methods. This method can reduce the receiver system complexity and is independent of the dummy sequence error. The change in TE should be considered because some of the subcarriers carry the dummy sequence (Ryu et al., 2004).

Hence, TE can be defined by

$$Transmission\ Efficiency\,(TE) = \frac{K}{K+L} \times 100\%$$

where K is the length of the signal and L is the length of the dummy sequence.

The DSI method outperforms C-SLM and O-PTS schemes in terms of BER performance; however, its PAPR performance will be inferior.

According to the literature, the most suitable application of DSI is in the WiMAX system where it does not require side information to be transmitted along with user data. However, there is a noticeable delay in finding the dummy sequence that affects the overall performance of the WiMAX system.

A faster DSI method is based on finding the dummy sequence without the need of iterative procedure (Chaokuntod et al., 2009; Uthansakul et al., 2009). The problem of an increased number of IFFTs is not seen in the DSI method. However, the complexity should be considered when designing the DSI method. As shown in Figure 4.20, the DSI method replaces the zeros of the signal with the dummy signals. If 55 zero slots of OFDM signal is replaced with dummy signals, according to Uthansakul et al. (2009), the bandwidth does not increase and so the efficiency has no degradation. However, the number of dummy signals should not exceed 55 due to the WiMAX standard (Chaokuntod et al., 2009). If more than 55 dummy signals are applied, the length of data and bandwidth will increase, and as a result, the TE will be degraded. The complexity of implementing the system will also be affected. The flowchart of the DSI algorithm is illustrated in Figure 4.20.

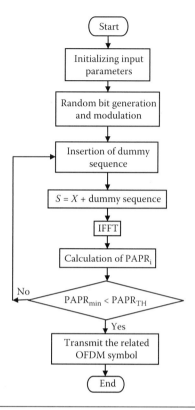

Figure 4.20   Flowchart of the DSI method.

As shown in the flowchart in Figure 4.20, the first step in the DSI method is to generate a random dummy sequence and perform the modulation process. Then, the dummy sequence is applied to the input signal. After IFFT, the PAPR is calculated and compared with a predefined threshold value. If the PAPR is less than the threshold, the signal is transmitted. Otherwise, the process of generating a random dummy sequence and applying it to the main signal is repeated until the calculated PAPR is less than the threshold. The performance of DSI is not attractive but the advantages of not adding complexity, TE degradation, and extra bits for side information show a good way for improvement. Most of the recent algorithms are based on a combination of DSI with other techniques (Kim et al., 2006; Varahram et al., 2010).

4.3.2.8 DSI-PTS   The block diagram of the DSI-PTS technique is shown in Figure 4.21. From this figure, the complex valued dummy signals are first generated and then added to the vector of data subcarriers.

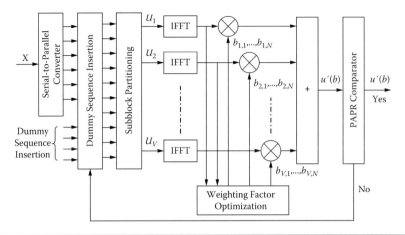

Figure 4.21 Block diagram of the proposed DSI-EPTS scheme.

The new vector in the frequency domain is then constructed from K-data and L-dummy subcarriers, respectively. L can be any number less than K. The new vector S is given by

$$S = [X_k, W_l] \qquad (4.16)$$

where $X_k = [X_{k,0}, X_{k,1}, ..., X_{k, N-L-1}]$, $k = 1,2, ..., K$ is the data subcarrier vector and $W_l = [W_{l,0}, W_{l,1}, ..., W_{l, L-1}]$, $l = 1,2, ..., L$ is the dummy signals vector.

After the generation of the optimum OFDM signal, then the PAPR is checked with the acceptable threshold that is predefined. If the PAPR value is less than the threshold, then the OFDM signal will be transmitted; otherwise the dummy sequence is generated again. This process is one iteration. The number of iterations can be increased to achieve the desired PAPR ($PAPR_{th}$) reduction but the processing time will also increase and cause the system performance to drop.

4.3.2.9 DSI-EPTS The block diagram of the DSI-EPTS technique is shown in Figure 4.21. From this figure, the complex valued dummy signals are first generated and then added to the vector of data subcarriers. The new vector in the frequency domain is then constructed from K-data and L-dummy subcarriers, respectively. L can be any number less than K. The new vector U is given by:

$$U = [X_k, W_l] \qquad (4.17)$$

where $X_k = [X_{k,0}, X_{k,1}, ..., X_{k, N-L-1}]$, $k = 1,2, ..., K$ is the data subcarrier vector and $W_l = [W_{l,0}, W_{l,1}, ..., W_{l, L-1}]$, $l = 1,2, ..., L$ is the dummy signals vector.

After the generation of the optimum OFDM signal, the PAPR is checked with the acceptable threshold that is predefined. If the PAPR value is less than the threshold, then the OFDM signal will be transmitted; otherwise the dummy sequence is generated again. This process is done in one iteration. The number of iterations can be increased to achieve the desired PAPR ($PAPR_{th}$) reduction, but the processing time will also increase and cause the system performance to drop. From the block diagram in Figure 4.21, X is the input signal with length N. The dummy sequence is added, which cause an increment in the IFFT length. Now the signal is partitioned into the V disjoint block $U_v = [U_1, U_2, ..., U_V]$ such that $\sum_{v=1}^{V} U_v = U$ and then these subblocks should be combined to minimize the PAPR in the time domain.

However, in the time domain the signal u_v is oversampled S times, which is obtained by taking an IFFT of length SN on U_v concatenated with $(S - 1)N$ zeros. After partitioning the signal and performing the IFFT for each part, then the phase factors $b_v = e^{j\phi_v}$, $v = 1, 2, ..., V$ are used for optimizing the U_v.

In time domain, the OFDM signal can be expressed as

$$u'(b) = \sum_{v=1}^{V} b_v u_v \qquad (4.18)$$

where $u'(b) = [u_0'(b), u_1'(b), ... u_{NS-1}'(b)]^T$. The objective is to find the optimum signal $u'(b)$ with the lowest PAPR. We should note that here, $N = K + L$, which means that there is no change in the length of the input signal after the addition of the dummy sequence. The subblock partition type that is applied here is based on interleaving, which is the best choice for PTS OFDM in terms of computational complexity reduction as compared to adjacent and pseudo-random; however, it has less PAPR reduction among them.

Then, the process is performed by choosing the optimization parameter $\tilde{b}$ with the following condition:

$$\tilde{b} = \arg\min\left(\max_{0 \le k \le NS-1} \left| \sum_{v=1}^{V} b_v u_{v,k} \right|\right) \qquad (4.19)$$

After finding the optimum $\tilde{b}$, then the optimum signal $u'(b)$ is transmitted to the next block. Now the PAPR of $u'(b)$ is checked whether it is in the range of the threshold PAPR ($PAPR_{th}$). After this additional task, the signal is transmitted, otherwise it is returned to the DSI block to generate the dummy sequence again. This process will continue until the PAPR is less than the $PAPR_{th}$.

To explain the DSI-PTS method (see Figure 4.22), consider L as the dummy sequence length, which later will be shown to be $L \le 55$ and N is the IFFT length, which is 256 in the case of a fixed WiMAX that includes 192 data carriers, 8 pilots, 55 zero paddings, and 1 DC subcarrier. Here, a complementary sequence is applied for the DSI (Ryu et al., 2004). X is the input signal with length N. After that the dummy sequence is added. The dummy sequence can be replaced with zeros in the data sample. This makes the IFFT length without change and decoding of the samples in receiver simpler. Now the signal is partitioned into the M disjoint block $S_m = [S_1, S_2, ..., S_M]$ such that $\sum_{m=1}^{M} S_m = S$ and then these subblocks should be combined

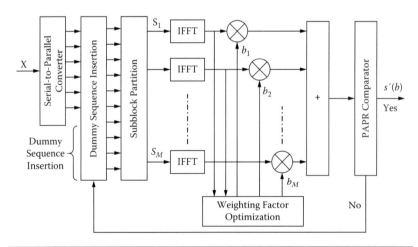

Figure 4.22   Block diagram of the DSI-PTS technique.

to minimize the PAPR in the time domain. However, in the time domain, the signal S_m is oversampled D times, which is obtained by taking an IFFT of length DN on X_m concatenated with $(D-1)N$ zeros. After partitioning the signal and performing the IFFT for each part, then the phase factors $b_m = e^{j\Phi_m}, m = 1, 2, ..., M$ are used for optimizing the S_m. In the time domain the OFDM signal can be expressed as

$$s'(b) = \sum_{m=1}^{M} b_m s_m \qquad (4.20)$$

where $s'(b) = [s'_0(b), s'_1(b), ...s'_{NF-1}(b)]^T$. The objective is to find the optimum signal $s'(b)$ with the lowest PAPR. We should note here that $N = K + L$, which means that there is no change in the length of the input signal after the addition of a dummy sequence. The subblock partition type that is applied here is based on interleaving, which is the best choice for PTS OFDM in terms of computational complexity reduction as compared to adjacent and pseudo-random, however it has less PAPR reduction.

Then the process is performed by choosing the optimization parameter $\tilde{b}$ with the following condition:

$$\tilde{b} = \arg \min \left(\max_{0 \le k \le NF-1} \left| \sum_{m=1}^{M} b_m s_{m,k} \right| \right) \qquad (4.21)$$

After finding the optimum $\tilde{b}$ then the optimum signal $s'(b)$ is transmitted to the next block. Now the PAPR of $s'(b)$ is checked to see whether it is in the range of the threshold PAPR ($PAPR_{th}$). After this additional task, the signal is transmitted; otherwise it is returned to the DSI block to generate the dummy sequence again. This process will continue until the PAPR is less than the $PAPR_{th}$.

4.3.3 A Discussion on the Current PAPR Reduction Solutions

Main PAPR reduction techniques have been discussed and compared in terms of data rate loss, distortion effect, transmitter complexity, receiver complexity, side information, and PAPR reduction performance. As shown in Table 4.1, the most promising methods are SLM and

Table 4.1 A Comparison of the PAPR Methods

MODIFIED SLM TECHNIQUES	PAPR IMPROVEMENT AT 10^{-4} CCDF	CCRR MUL./ADD.	NUMBER OF IFFTs
Jeon et al., 2011	4 dB	35%/50%	16
Naeiny and Marvasti, 2011	3.8 dB	—	16
Hong and Har, 2010	2.6 dB	63%/69%	8
Wang et al., 2011	2.7 dB	14%/22%	8
Kim et al., 2006	3.1 dB	—	8

PTS, which have a high PAPR performance without distortion effects. Therefore, many recent studies have focused on modifications to the SLM and PTS methods.

In terms of complexity reduction, the maximum achievement is about 63% in the number of addition and 69% in the number of multiplications (Hong and Har, 2010); however, the PAPR performance is not attractive.

Therefore, according to our literature review, there is no algorithm to reduce the PAPR of the OFDM signal with low complexity.

In order to reduce PAPR with less complexity, two methods have been designed. One of the proposed methods is a combination method using a modified version of DSI and SLM methods.

4.4 Design of the Proposed DSI-SLM Scheme

The main PAPR reduction solutions have been reviewed. In this section, a new PAPR reduction scheme called DSI-SLM is proposed, which is based on a combination of modified DSI with a conventional SLM (C-SLM) method. Since the SLM method outperforms other techniques and because the DSI method reduces the PAPR without adding any complexity or extra bits to the system, the combination of these two techniques is proposed, which is expected to have significant PAPR performance compared to all previous methods.

In the following sections, DSI and C-SLM are separately simulated and the proposed DSI-SLM scheme is applied to a typical OFDM signal, and its PAPR performance is evaluated and compared with other methods. The DSI-SLM scheme is designed to enhance the drawbacks of the previous methods. The modified receiver design of the DSI-SLM method is also simulated to verify the BER performance.

4.4.1 *The Proposed DSI-SLM Scheme*

DSI and SLM methods have some practical limitations. In the DSI method, the addition of dummy signals increases the length of the signal and therefore it degrades the transmission efficiency. The C-SLM method requires a high number of IFFT blocks to achieve the desired PAPR reduction, which leads to a higher number of side information bits and high complexity.

The proposed DSI-SLM combines a modified dummy sequence based on the DSI algorithm with a random phase sequence developed from the SLM algorithm.

The architecture of the DSI-SLM is presented in Figure 4.23 and it is designed based on the following mathematical derivations. Referring to Figure 4.23, following the serial-to-parallel block, a random dummy sequence is inserted into the input WiMAX signal denoted X. The following equation shows mathematically how a dummy sequence with a length of L is added to the input signal X with length K and signal S with length $K + L$ is generated:

$$S = MUX(X,D), \quad L = 10, 30, \text{ and } 55 \qquad (4.22)$$

where MUX is the multiplex function of two vectors X and D, which have the lengths K and L, respectively, and the length of S is $K + L$.

As long as the OFDM signal is based on the IEEE 802.16d standard or fixed WiMAX, K is equal to 201. If the IFFT length becomes 512 or 1024 based on the IEEE 802.16e standard or WiMAX mobile signal, the value of K will be modified. The OFDM signal based on

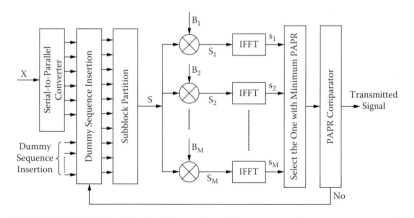

Figure 4.23   Block diagram of the DSI-SLM transmitter.

fixed WiMAX has 256 subcarriers with 55 zeros, which are replaced with a dummy sequence, thus the maximum value of L is 55.

In simulations, L indicates the length of the dummy sequence. Several simulations are performed to study the effects of different values of L (L = 10, 33, and 55). Since there are 55 zero slots in the OFDM signal frame, the maximum number of dummies that can be applied without TE degradation is 55. Here the simulation is performed for 10 and 30 dummies as well, and it is shown that by increasing the number of dummies, the PAPR reduction can be improved. However, in order to avoid spectrum reshaping, occupying all the zero slots is not advised by some researchers. It should be noted that the dummy sequences can also be added to the OFDM symbol, which results in an increasing of signal bandwidth and degradation of TE. According to Uthansakul et al. (2009), replacing 55 zero slots of the OFDM symbol with the dummy sequences has no degradation on the spectrum efficiency, therefore, the dummy sequence is inserted within 55 slots of the OFDM symbol of the WiMAX signal.

In the SLM section, M random phase rotations (B_1, B_2, ..., B_M) with length of $K + L$ are multiplied to the copies of the original signal (S) to produce candidate signals. The procedure can be given by

$$
\begin{bmatrix} S_i(1) \\ S_i(2) \\ \vdots \\ S_i(K+L) \end{bmatrix} = S \times \begin{bmatrix} B_i(1) \\ B_i(2) \\ \vdots \\ B_i(K+L) \end{bmatrix}, i = 1, 2, ..., M \qquad (4.23)
$$

Then the IFFT is performed on M sequences as given by

$$
S = \begin{bmatrix} s_1 \\ s_2 \\ \vdots \\ s_M \end{bmatrix}^T = \frac{1}{N} \begin{bmatrix} \sum_{K=0}^{N-1} S_1(K)\exp(2\pi\, jnk/N) \\ \sum_{K=0}^{N-1} S_2(K)\exp(2\pi\, jnk/N) \\ \vdots \\ \sum_{K=0}^{N-1} S_M(K)\exp(2\pi\, jnk/N) \end{bmatrix}^T \qquad (4.24)
$$

where T is the vector transpose and N is the OFDM length. It should be noted that the length of IFFT (N) is equal to $K + L$.

Following IFFT, the PAPRs of vector $\{s_1, s_2, ..., s_M\}$ are computed and the sequence with the lowest PAPR is selected. The optimum PAPR can be calculated as follows:

$$PAPR_{opt} = min \{ PAPR[s]\} = min\{PAPR[s_1, s_2, ..., s_M]\} \quad (4.25)$$

The $PAPR_{opt}$ is compared with a predefined threshold, $PAPR_{th}$, as shown in Equation (4.26). If the $PAPR_{opt}$ is lower than the $PAPR_{th}$, the sequence with the minimum PAPR will be transmitted; otherwise, it gives feedback to generate a new dummy sequence.

$$PAPR_{opt} \langle PAPR_{th} \quad (4.26)$$

Another dummy sequence is inserted and the PAPR is calculated and compared with the threshold again. As an example, if the computed PAPR for 8 candidate signals ($M = 8$) is $PAPR_{(S)} = (8.5, 7.2, 5.9, 6.9, 8, 6.5, 8.5, 7.9)$ dB and the threshold PAPR is $PAPR_{th} = 6$ dB, then the $PAPR_{opt}$ becomes 5.9 dB. Because the optimum PAPR is less than the threshold PAPR, the sequence with the optimum PAPR, which in this case is the third sequence, S_3, will be transmitted; otherwise the minimum PAPR among the candidate signals will be transmitted.

The other form of the proposed method is to apply dummy signals after phase sequence multiplication, exactly before the IFFT blocks, but this results in increasing the complexity as this process should be performed for each partition. Simulation results show that increasing M leads to a more significant PAPR reduction, but a higher number of IFFT blocks is required to generate candidate signals and so according to the following equation, the higher number of bits should be sent to the receiver to indicate which one is selected:

$$\text{Number of side information bits} = log_2{M}$$

The flowchart of the proposed DSI-SLM scheme is presented in Figure 4.24. In the first step, the parameters of the simulation are defined, such as iteration, the number of dummy signals, and the value of the threshold PAPR. At the end, the optimized sequence is given to the transmission stage and also the channel. Then the receiver can receive the signal. The receiver of the DSI-SLM scheme is also designed as shown in Figure 4.25, in which the received signal

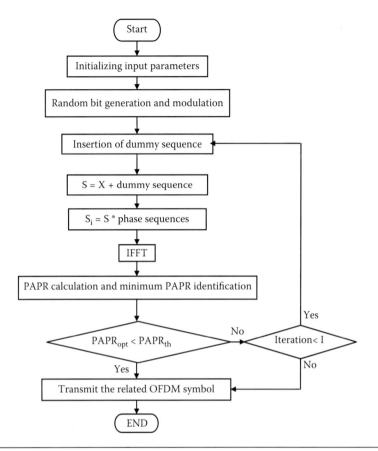

Figure 4.24   Flowchart for the proposed DSI-SLM scheme.

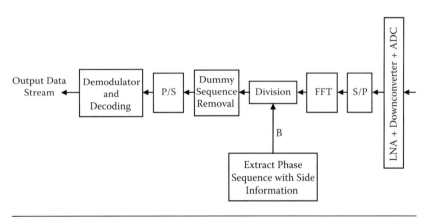

Figure 4.25   Block diagram of the DSI-SLM receiver.

is downconverted and after serial-to-parallel conversion, it reaches the FFT block.

The FFT algorithm is based on Equation (4.27) (Schniter, 2004) as follows:

$$X(k) = \sum_{n=1}^{N-1} x(n) e^{-j2\pi kn/N}, \ k = 0, ..., N-1 \qquad (4.27)$$

where FFT length (N) is equal to the total length of the signal, which according to Equation (4.27) is $K + L$. The side information is extracted from the received signal and so the multiplied phase sequence is detected. In the receiver, the inverse operation of the multiplication is required, which is shown by division in the receiver block diagram of Figure 4.25.

Following the FFT, the received signal s_i should be divided by the detected phase sequence as shown by B_i in Equation (4.28). The quotient of this complex division will be the signal S that includes the dummy sequence, as shown by

$$\hat{S} = \frac{\hat{S}_i}{\hat{B}_i} \qquad (4.28)$$

To have the pure signal that contains the original data, elimination of the dummy sequence is performed. Thus, this process can be executed by a subtraction operation, which is shown by

$$\begin{bmatrix} X(1) \\ X(2) \\ \vdots \\ X(K) \end{bmatrix} = DEMUX(S, D), \ L = 10, 30, \text{ and } 55 \qquad (4.29)$$

where $DEMUX$ is the demultiplexing function of the two vectors S and D. By discarding the dummy signal the original signal can be retrieved.

Signal X needs to be demodulated and decoded after the parallel-to-serial converter and the original data stream is extracted. The flowchart of the receiver simulation program is shown in Figure 4.26.

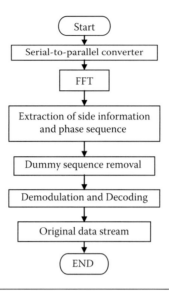

Figure 4.26 Flowchart for the DSI-SLM receiver.

The DSI-SLM scheme has many advantages compared to the other PAPR reduction techniques. One advantage is that unlike clipping and companding, the DSI-SLM scheme does not exhibit any distortion into the signal spectrum. The reason is that the DSI-SLM scheme changes the probability of the PAPR and unlike the clipping method, DSI-SLM does not modify the envelope of the signal in time domain. The other advantage of DSI-SLM over the other methods, such as tone injection and tone reservation, is the simplicity of the transmitter design. The main advantage of DSI-SLM compared to conventional SLM is the significant reduction in complexity while the PAPR performance can be the same. The main drawback of the DSI-SLM method is the reduction of spectrum efficiency due to the side information and the dummy sequence insertion. Although the complexity is significantly reduced by applying the DSI-SLM method, this is achieved by inserting more dummy signals, which degrade the spectrum efficiency. Hence the other PAPR method is required to overcome this problem.

The transmission efficiency (TE) changes as expressed by

$$Transmission\ Efficiency\ (TE) = \frac{K}{K+L} \times 100\% \qquad (4.30)$$

where K is the total number of subcarriers and L is the number of dummy sequences. In DSI-SLM, since dummy sequences are located in 55 zero guard subcarriers of OFDM signal, the total number of subcarriers remains the same, which yield TE 100%.

The side information is an important aspect in studying PAPR reduction techniques that involves transmitting some extra bits to inform the receiver about the modifications that have been done on the signal in the transmitter in order to extract the original signal. One way is that the side information can be transmitted using a separate channel at the expense of spectrum efficiency. Spectrum efficiency or bandwidth efficiency refers to the information rate that can be transmitted over a given bandwidth. As mentioned earlier, the number of required side information bits in C-SLM is $log_2 M$ where M is the number of candidate signals (Baxley and Zhou, 2007; Ghassemi and Gulliver, 2009; Le Goff et al., 2009).

In the DSI-SLM scheme, the required side information also can be calculated from the aforementioned formula. It should be noted that the addition of dummy sequences does not affect the side information, because the dummy signals are discarded at the receiver from the end of the OFDM signal. In the DSI-SLM scheme, according to the simulation results, fewer candidate signals (M) are required to gain the comparable PAPR performance. While M is reduced, fewer bits are required to be transmitted as side information. This means that the spectrum efficiency will be enhanced in the DSI-SLM scheme.

4.4.2 DSI-SLM Computational Complexity

There are some considerations for practical implementation of the DSI-SLM system. The computational complexity is one of them that reflects the hardware resource requirements. The IFFT block is the main hardware resource consumer block in implementing OFDM systems (Baxley and Zhou, 2007; Hou et al., 2011; Muller and Huber, 1997c). The IFFT changes the distribution of the signal without alteration of its average power. So, the number of IFFT blocks that is required for implementation defines the computational complexity of a PAPR reduction method. To evaluate the performance of the DSI-SLM, the OFDM signal with $N = 256$ subcarriers based on a WiMAX fixed line standard is used in the simulations.

When the number of carriers is $N = 2^n$, the number of multiplications (n_{mul}) and the number of additions (n_{add}) of both the C-SLM and DSI-SLM techniques are given by

$$n_{mul} = 2^{n-1} nM$$

$$n_{add} = 2^n nM$$

(4.31)

where M is the number of candidate signals. For the proposed DSI-SLM, $M = 2$, which is due to the comparable PAPR performance with C-SLM when $M = 8$.

In terms of computational complexity, the DSI-SLM scheme has significant improvement. According to Hou et al. (2011), the PAPR improvement is measured by using the CCRR (which was introduced earlier):

$$CCRR = \left(1 - \frac{Complexity\ of\ the\ DSI\text{-}SLM}{Complexity\ of\ the\ C\text{-}SLM} \right) \times 100\% \quad (4.32)$$

If the proposed scheme performs with the same complexity as the C-SLM technique, the CCRR will be 0%. If the new method has less complexity, CCRR will have a higher value. The complexity of the proposed method of DSI-SLM is calculated and the CCRR of the proposed DSI-SLM over the C-SLM method is given in Table 4.2.

The value of M is 2 for the proposed scheme when comparing with the C-SLM with $M = 2$, 4, and 8, but when the DSI-SLM is compared with the C-SLM method with $M = 16$, the value of M has to be 4 to have fair comparison. The reason is that the PAPR reduction

Table 4.2 Computational Complexity of the DSI-SLM and the Conventional SLM Methods

COMPLEX COMPUTATION	NUMBER OF SUBBLOCKS	CONVENTIONAL SLM	DSI-SLM	CCRR
Multiplication	$M = 2$	2048	2048 ($M = 2$)	0%
Addition	$M = 2$	4096	4096 ($M = 2$)	0%
Multiplication	$M = 4$	4096	2048 ($M = 2$)	50%
Addition	$M = 4$	8192	4096 ($M = 2$)	50%
Multiplication	$M = 8$	8192	2048 ($M = 2$)	75%
Addition	$M = 8$	16384	4096 ($M = 2$)	75%
Multiplication	$M = 16$	16384	4096 ($M = 4$)	75%
Addition	$M = 16$	32768	8192 ($M = 4$)	75%

of the DSI-SLM scheme with $M = 2$ is comparable with the C-SLM when $M = 2$, 4, and 8; however when $M = 16$, the DSI-SLM with $M = 2$ does not have better PAPR or the same PAPR performance, so in comparison with the computational complexity, it is better to consider the DSI-SLM with $M = 4$, which is compared with C-SLM when $M = 16$ as shown in Figure 4.26.

In the first two rows of Table 4.2 for $M = 2$, the reduction is 0%, because in both DSI-SLM and C-SLM, the number of multiplications and additions are the same, so the complexity of the proposed scheme is the same as the conventional method.

When $M = 4$, the number of IFFTs of the DSI-SLM is still 2, so from Equation (4.32), CCRR is 50%. When $M = 8$, this improvement is 75% and the maximum improvement in terms of complexity is achieved when $M = 16$. The CCRR of the proposed DSI-SLM over the conventional SLM method is 75% when $M = 16$. In this condition, the proposed scheme can achieve the same amount of PAPR reduction with improved complexity by 75%.

The total complexity of the C-SLM method when oversampling factor $F = 1$ is given by Equation (4.33). In order to calculate the total complexity the number of multiplications and additions required for generating candidate signals, IFFT calculation and computing PAPR have to be considered:

$$T_{C-SLM} = 5MN\log N + 12\,MN - 6N \qquad (4.33)$$

To calculate the total complexity of C-SLM, $N(M{-}1)$ complex multiplication is required to create X_M. Next, M length K IFFT are needed to generate x_n. Each IFFT requires $N/2\log N + N/2$ complex multiplication and $N\log N$ complex addition. Finally, to calculate the PAPR, the absolute of the magnitude of x_n must be calculated at each n, which comes at the expense of $2MN$ real multiplications and MN real additions. In general, a complex multiplication takes four real multiplications and two real additions. Hence the total number of multiplications and additions of the C-SLM are given by

$$M_{C-SLM} = 2MNF(\log N + 2) + 4N\,(M - 1) \qquad (4.34)$$

$$A_{C-SLM} = MNF(3\log N + 2) + 2N\,(M - 1) \qquad (4.35)$$

where F is the oversampling factor.

For the DSI-SLM scheme, this value when $M = 2$ can be given by

$$T_{DSI-SLM} = T_{C-SLM} + 2L = 5MN\log N + 12MN - 6N + 2L \quad (4.36)$$

where M is the number of candidate signals, N is the number of sub-carriers, and L is the dummy sequence length. In Equation (4.36), the number of complex additions for applying dummy sequences should be also considered, which is twice the number of real additions and is presented by $2L$ in this equation. To calculate $T_{DSI-SLM}$, we apply Equation (4.26) with different values of M.

According to Table 4.3, the DSI-SLM achieves 76% improvement in total computational complexity. It can be observed that when $M = 2$ for both methods, the CCRR becomes negative due to the addition of dummy sequences, according to Equation (4.36). For other values of M, the CCRR shows significant improvement. It should be noted that the value of M in DSI-SLM is comparable with C-SLM according to the results shown later in this section.

Table 4.4 presents the CCRR comparison of the DSI-SLM scheme with two other methods: DSI-PTS and modified SLM (Li et al., 2010; Varahram et al., 2010). The complexity of the modified SLM (Li et al., 2010) is based on the algorithm proposed to replace IFFTs and does not consider the total computational complexity. Therefore, the complexity

Table 4.3 Total Computational Complexity Reduction Comparison between C-SLM and DSI-SLM

C-SLM	DSI-SLM	$CCRR_{DSI-SLM}$
10773 ($M = 2$)	10883 ($M = 2$)	−1%
23082 ($M = 4$)	10883 ($M = 2$)	53%
47700 ($M = 8$)	10883 ($M = 2$)	77%
96936 ($M = 16$)	23192 ($M = 4$)	76%

Table 4.4 Total Computational Complexity Reduction Comparison between DSI-PTS and DSI-SLM

M	$T_{DSI-PTS}$	$T*$	$T_{DSI-SLM}$
2	5075	10773	10883
4	132975	23082	23192
8	67112615	47700	47810

* From Li, Ch. P., Wang, S. H., and Wang, Ch. L. 2010. *IEEE Transactions on Signal Processing* 58(5):2916–2921.

of the modified SLM in the second column of Table 4.4 is slightly less than DSI-SLM; however its total complexity will become higher.

According to Baxley and Zhou (2007), when $M = 2$ the PTS method has less complexity compared to the SLM method; therefore the total complexity of DSI-PTS is predicted to be less than DSI-SLM in this condition. However, when M is increased to 4 and 8, the complexity of DSI-PTS is increased. This proves that the DSI-SLM scheme outperforms the other PAPR reduction methods in terms of complexity and hardware resource consumption.

4.5 Simulation Results and Analysis

Following the theoretical design, the DSI-SLM scheme is simulated by using MATLAB simulation tools. Its m.file code is provided in Appendix B. Random data bits based on QPSK modulation are generated and stored in matrices.

The DSI-SLM is performed for several iterations and the output signals are captured and stored in matrices separately and their PAPRs are calculated. Then, the sequence with minimum PAPR is selected for transmission.

Here, the CCDF of the C-SLM, DSI, and DSI-SLM schemes are compared together. The PAPR performance of the C-SLM method is shown in Figure 4.27, which shows that the C-SLM method is simulated for different numbers of candidate signals ($M = 2, 4, 8,$ and 16). As shown by the unmarked solid line, the original OFDM signal has about 11.8 dB PAPR at CCDF = 10^{-4} = 0.01%. As shown by the circle-marked curve, when the C-SLM method with two candidate signals ($M = 2$) is applied, the PAPR is reduced to about 10.4 dB.

When the C-SLM method with four candidate signals ($M = 4$) is applied, the PAPR is decreased to about 9.2 dB. When the number of candidate signals is increased to 8 and 16, the PAPR is reduced to 8.4 dB and 7.9 dB at CCDF = 0.01%, respectively.

The reason is that when M is increasing, the candidate signals are more likely to generate less PAPR, thus the minimum PAPR will be less than before.

The CCDF result of the C-SLM method with $N = 1024$ is shown in Figure 4.28. By comparing this figure with Figure 4.27, it is obtained that the PAPR is increased. In Figure 4.29, the original OFDM signal has about 12.5 dB PAPR at 10^{-4} CCDF, which is reduced to about 10.9 dB, 10 dB, 9.3 dB, and 8.8 dB when M is 2, 4, 8, and 16, respectively. It can be concluded that the C-SLM method with $M = 16$ and $N = 256$ has reduced PAPR by about 4 dB and the C-SLM method with $M = 16$, $N = 1024$ has reduced PAPR by about 3.7 dB, which

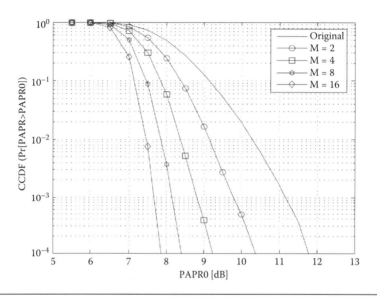

Figure 4.27   CCDF of the C-SLM method ($N = 256$).

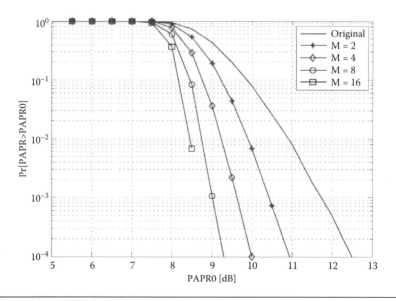

Figure 4.28   CCDF of the C-SLM method ($N = 1024$).

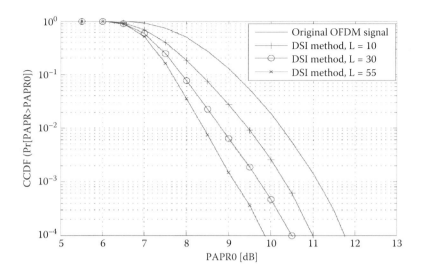

Figure 4.29 CCDF of the DSI method.

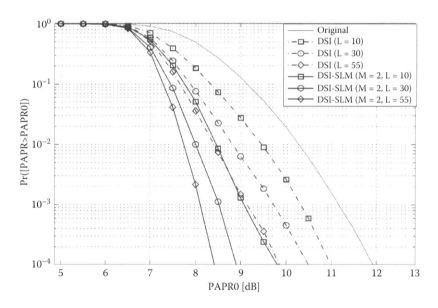

Figure 4.30 CCDF comparison of the DSI method and DSI-SLM scheme.

shows slight degradation. The reason is that by increasing the number of subcarriers (N), the probability of having high PAPR values is increased.

Figure 4.30 shows the PAPR performance of the DSI method when the number of iterations in the DSI loop is 10. The PAPR of

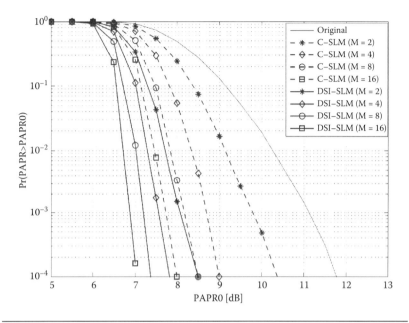

Figure 4.31 CCDF comparison between the DSI-SLM scheme and the C-SLM method (N = 256).

the original OFDM signal is about 11.8 dB at CCDF = 0.01% as shown by the unmarked curve in Figure 4.30.

When the DSI method with L = 10 is applied, the PAPR is reduced to 11 dB. When the length of the dummy sequence L is increased to 30 and 55, the PAPR is reduced to about 10.5 dB and 9.8 dB at CCDF = 0.01%, respectively. This result shows that by increasing the number of dummy signals L, the PAPR reduction is improved; however the PAPR performance depends on the number of DSI iterations (I). It should be noted that when the number of iterations is increasing, the PAPR reduction will not be significant due to the limited capability of the DSI method.

Figure 4.31 shows the comparison between the DSI method and DSI-SLM scheme when the length of the dummy sequence (L) is 10, 30, and 55, respectively. As shown in Figure 4.31, the PAPR of the original OFDM signal without any PAPR reduction is about 12 dB at CCDF = 0.01%. When the DSI method is applied, as shown by dash-dot curves, by increasing the length of the dummy sequence from 10 to 30 and 55, the PAPR is reduced to about 11 dB, 10.5 dB, and 9.8 dB, respectively.

As shown by the solid line curves in Figure 4.31, by increasing the dummy length (L) of DSI-SLM from 10 to 30 and 55, the PAPR is reduced to about 9.8 dB, 8.9 dB, and 8.4 dB, respectively. The PAPR improvement is about 1.2 dB, 1.4 dB, and 1.6 dB when L is 10, 30, and 55, respectively. This means that an average improvement of 1.4 dB in PAPR performance is achieved.

The proposed DSI-SLM scheme is also compared with C-SLM with various values of M (2, 4, 8, and 16). Figure 4.32 presents the CCDF comparison between DSI-SLM and the C-SLM when L in DSI-SLM is fixed and equal to 55. As shown in Figure 4.32, the PAPR of the original OFDM signal is about 11.8 dB at CCDF = 0.01%. When the C-SLM method with various numbers of candidate signals (M = 2, 4, 8, and 16) is applied, the PAPR is reduced to about 10.4 dB, 9 dB, 8.5 dB, and 8 dB, respectively.

When the DSI-SLM scheme with the same amount of candidate signals, which indicates the number of IFFT blocks (M = 2, 4, 8, and 16), is applied, the PAPR of the OFDM signal is reduced to about 8.5 dB, 7.8 dB, 7.4 dB, and 7 dB, respectively. When M is 2, 4, 8, and 16, about 1.8 dB, 1.2 dB, 1.2 dB, and 1 dB reduction in PAPR is achieved, respectively. Therefore it can be concluded that within the range of M = 1 to 16, an average improvement of 1 dB in PAPR reduction performance is achieved.

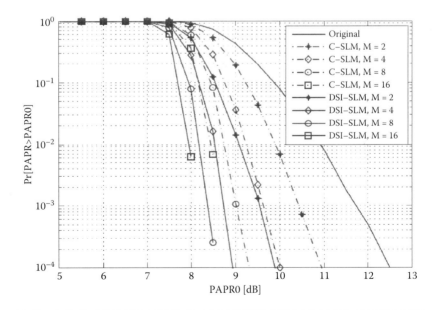

Figure 4.32   CCDF comparison between the DSI-SLM and C-SLM techniques (N = 1024, L = 55).

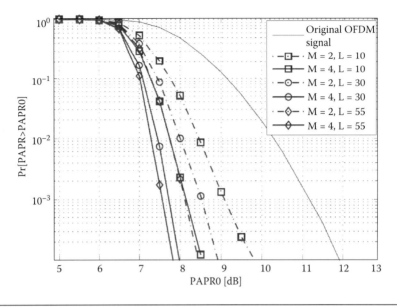

Figure 4.33 CCDF of the DSI-SLM scheme with various values of M and L ($N = 256$).

As shown in Figure 4.33, the PAPR reduction in C-SLM when M is 8 and 16 is very close to the DSI-SLM with M equal to 2 and 4. This means that in practice, a fewer number of IFFT blocks are required to be implemented to achieve the same amount of PAPR reduction with the same OFDM signal, which explains the 76% improvement in computational complexity that was presented in Table 4.2.

As shown in Figure 4.33, when the number of subcarriers (N) is increased from 256 to 1024, the original OFDM signal has about 12.5 dB PAPR at 10^{-4} CCDF.

The PAPR reduction of C-SLM and DSI-SLM are compared in Figure 4.33, which shows that DSI-SLM reduces the PAPR by about 1 dB more than C-SLM. This is similar to the result of $N = 256$ as shown in Figure 4.32.

Figure 4.34 shows the CCDF comparison of the DSI-SLM scheme with various values of L and M. The dash-dot curves show the PAPR performance of the DSI-SLM when $L = 10$, 30, and 55 and $M = 2$ and solid lines show the result when $L = 10$, 30, and 55 and $M = 4$. As shown in Figure 4.34, when L is fixed and M is increased, an average improvement of 1 dB can be achieved. When M is fixed and L is increased, an average improvement of 0.4 dB can be achieved.

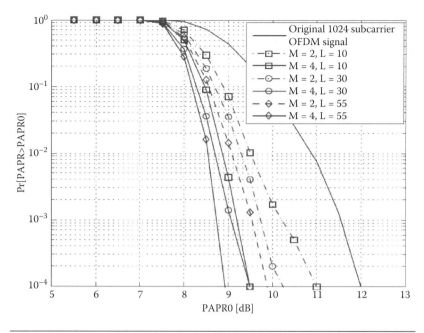

Figure 4.34 CCDF of the DSI-SLM scheme with various values of M and L ($N = 1024$).

For real-time implementation, the computational complexity and TE of each condition also should be considered.

As mentioned before, by increasing M more hardware resources is required and by increasing L a higher signal bandwidth is necessary.

The performance of DSI-SLM with $N = 1024$ and various values of M and L is presented in Figure 4.34. Similar to Figure 4.33, the performance is degraded due to increasing the number of subcarriers, however, the PAPR reduction can be improved by increasing L and M to 55 and 16, respectively.

In Varahram et al. (2010), the DSI-PTS scheme is proposed, which is a combination of the PTS scheme and DSI method. Here, the DSI-SLM scheme is also compared with the DSI-PTS method. Figure 4.35 shows the PAPR performance comparison of these two techniques.

The length of the dummy sequence (L) in both DSI-SLM and DSI-PTS techniques is fixed to 55 and two conditions of $M = 2$ and $M = 4$ are examined for both DSI-SLM and DSI-PTS methods. The PAPR performance of the DSI-SLM outperforms DSI-PTS while the complexity

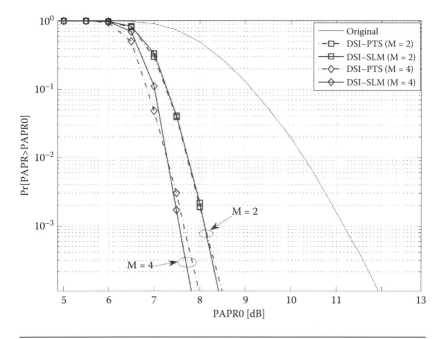

Figure 4.35 CCDF comparison of DSI-PTS and DSI-SLM (N = 256).

of DSI-PTS is inferior. The reason is that the complexity of PTS as shown in Varahram et al. (2010) is more than SLM due to the exhaustive searching for optimum phase sequence except $M = 2$, while in SLM the searching operation does not exist. Hence, the DSI-SLM scheme outperforms the DSI-PTS technique in terms of PAPR and complexity.

As shown in Figure 4.36, when the number of subcarriers is increased to 1024, the comparison between the PAPR performance of the DSI-SLM scheme and DSI-PTS shows about 0.1 dB difference between the DSI-SLM and DSI-PTS method.

According to the algorithm, the PAPR performance of the DSI-SLM scheme can be improved by increasing the number of iterations in the DSI feedback loop. The reason is that when the number of DSI iterations (I) is increased and if the PAPR is not less than the threshold, the DSI loop is repeated for more times and as a result more various dummies are generated. Therefore, the PAPR of more various candidate signals are measured and compared with the threshold. Hence, the probability of having lower PAPR is increased. All the presented results are based on 10 iterations of DSI ($S = 10$). Now the program is tested for more iterations. Figure 4.37 shows the PAPR

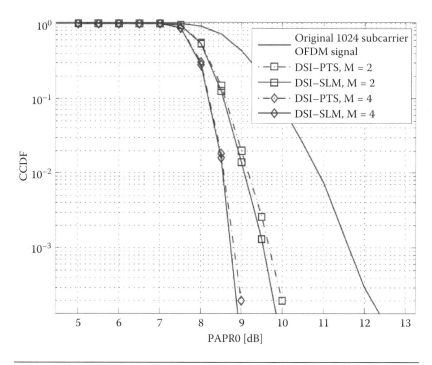

Figure 4.36   CCDF comparison of DSI-PTS and DSI-SLM ($N = 1024$).

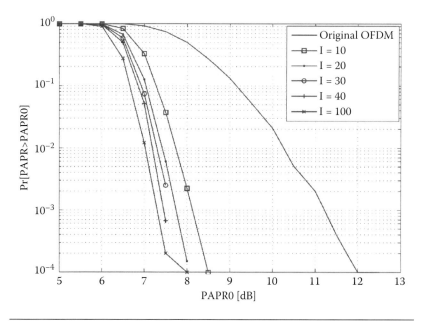

Figure 4.37   PAPR performance of DSI-SLM for various numbers of DSI iterations (I).

performance comparison of DSI-SLM when $M = 2$ and the number of iterations for DSI is varied ($S = 10, 20, 30, 40,$ and 100).

According to Figure 4.37, when $S = 10, 20, 30, 40,$ and 100, the PAPR of 8.5 dB, 8 dB, 7.8 dB, 7.6 dB, and 7.5 dB are achieved, respectively. It is shown that by increasing the number of DSI iterations (I), the PAPR performance is enhanced. It is always a trade-off between the PAPR performance and the processing delay, which causes data rate loss. While a higher number of iterations leads to higher PAPR performance, the data rate will be degraded. It should be noted that by increasing the number of iterations, the improvement of PAPR reduction is not significant.

In order to evaluate BER performance of the DSI-SLM scheme, the additive white Gaussian noise (AWGN) channel and the receiver are simulated. According to Kumar et al. (2007), the AWGN channel is a sufficient model for studying the BER of OFDM signals.

Here, in Figure 4.38, the BER performance comparison of the C-SLM and the proposed DSI-SLM scheme is shown when the AWGN channel is applied and $M = 2$ and $M = 4$. It is evident that the BER has a slight degradation for both $M = 2$ and $M = 4$ when the proposed DSI-SLM scheme is applied compared to the C-SLM method.

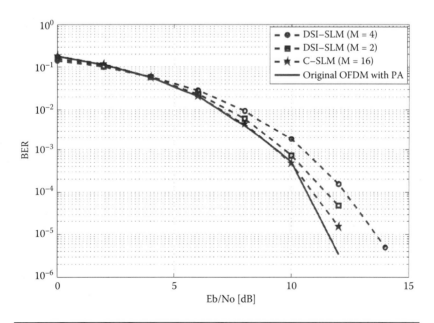

Figure 4.38 Comparison of the BER performance of DSI-SLM and C-SLM.

However, the BER performance is still acceptable according to Zhang and Thibault (2010), in which to provide spectrum efficient transmission, the BER should be below 10^{-4} at $S_b/N_b = 14$ dB.

The reason of degradation is that when the signal is modified in the transmitter, the probability of receiving a signal with corrupted bits in the receiver is increased, which means that the BER is degraded.

4.6 Results Discussion

According to the simulation result comparison of the proposed DSI-SLM with DSI, C-SLM, and DSI-PTS methods, it is seen that the DSI method has weak performance but when it is combined with an SLM-based technique, the PAPR performance and complexity are enhanced. Generating dummy signals and adding it to the original signal created a situation that caused the PAPR to become low. This situation requires some iteration to be applied to achieve better PAPR reduction. It should be noted that due to the random nature of the IFFT to create an OFDM signal, the condition to have minimum PAPR can only be met while several iterations are applied and there is no condition to know the accurate number of iterations.

On the other hand, the C-SLM method has very high complexity due to the high number of IFFTs and complex multiplications, but when it is combined with DSI, it can achieve better PAPR reduction with less complexity.

As shown in Figure 4.39, the PAPR performance of the DSI-SLM is close to the DSI-PTS method, but since the PTS alone has higher complexity than the SLM method, the complexity of the DSI-PTS is also higher than the DSI-SLM scheme. The PAPR performance of the DSI-SLM scheme can be enhanced by increasing the number of dummy sequences at the expense of bandwidth and spectrum efficiency. As shown by simulation results, increasing the dummy sequences has a direct effect on PAPR performance in which a higher number of dummy sequences improves the PAPR performance. However, increasing the number of dummy sequence degrades the spectrum efficiency and also causes data rate loss. Here, the zeros of the OFDM symbol are replaced with dummy sequences to maintain the spectrum efficiency and data rate loss. Another important point during analyzing the proposed scheme is

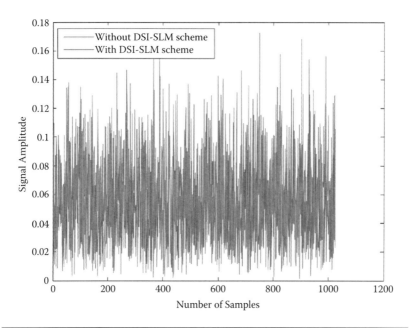

Figure 4.39 The OFDM signal before and after DSI-SLM.

the side information, which also degrades the spectrum efficiency. As discussed before, in the DSI-SLM scheme the PAPR of $M = 2$ is near to the C-SLM result with $M = 8$ and $M = 16$. This means that in the DSI-SLM scheme, only one bit of side information is required.

When this bit is zero, it indicates that the first candidate signal is transmitted and when it is one, it means that the second candidate signal is transmitted. In C-SLM with comparable results more than 3 bits are required for side information. However, the side information is still a drawback for the DSI-SLM scheme and so the robust method needs to be investigated, which will be introduced in the next chapter.

In some studies, the PAPR performance is studied using time domain symbols. Figure 4.39 presents 1024 samples of the output signal with and without PAPR. The light colored lines indicate samples of the output signal without PAPR reduction, and the dark colored lines indicate samples of the output signal when PAPR reduction is applied. It can be observed that the OFDM signal peaks are suppressed. However, the reduction seems to be insignificant because the DSI-SLM scheme is a probabilistic method and the reduction is based on signal modification, therefore, the time domain graph is not an accurate study tool for this case.

4.7 The Optimum Phase Sequence with the Dummy Sequence Insertion Scheme

In this section, the second proposed method called the optimum phase sequence with dummy sequence insertion (OPS-DSI) is introduced. As discussed earlier, the proposed DSI-SM scheme enhanced the PAPR performance while the number of IFFTs is reduced from 16 to 2 and 4 compared to C-SLM. In this section, the proposed OPS-DSI scheme only needs one IFFT while its PAPR performance is even better compared with DSI-SLM. The flexibility of the OPS-DSI is increased by controlling the number of iterations and trade-off between data rate and PAPR performance in which higher number of iterations results in better PAPR performance at the expense of data rate loss while a lower iteration number exhibits less PAPR reduction with less data rate degradation. By applying the OPS-DSI scheme, the bandwidth efficiency will not be degraded due to any side information bits. Side information can be replaced with the dummy sequence bits and be transmitted with the signal. The PAPR performance and complexity of the OPS-DSI will be compared with the DSI-SLM to verify the proposed scheme.

4.7.1 Design of the OPS-DSI Scheme

As explained earlier, the PAPR performance can be controlled with the DSI iteration loop. When the number of iterations of the DSI loop is increased, the PAPR performance is enhanced but the simulation program takes more time to be performed, which results in a time delay and data rate that should be considered in practical scenarios. The main idea behind the OPS-DSI scheme is to extend the control to the phase sequence generation and increase the flexibility of the system by having two loops for the dummy generation and phase sequence multiplication. When the phase sequence multiplication is controlled with the number of iterations, since the PAPR is compared with a threshold value, it means that the optimum phase sequence will be selected in order to be multiplied with the input signal to achieve minimum PAPR reduction. Another main concept in OPS-DSI is to have dummy insertion after the process of phase sequence multiplication. In this way, the index of phase sequence can be inserted within the dummy sequence and thus the problem of side information can be solved. Figure 4.40 shows the block diagram of an OPS-DSI scheme.

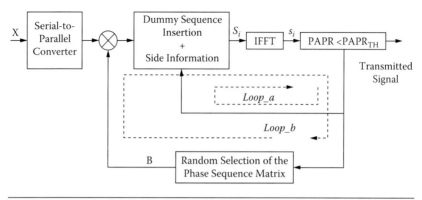

Figure 4.40 Block diagram of the OPS-DSI scheme, transmitter.

The input signal denoted by X is passed through the serial-to-parallel converter and then X is multiplied with a random phase sequence, B, selected from a matrix that is stored in memory. Following the multiplication, the dummy sequence is inserted into the input signal. The dummy sequence is also randomly generated from the 4 bits index of matrix B. The PAPR is computed and compared with a threshold. There are two loops in the OPS-DSI algorithm. If the PAPR is not less than the threshold, *Loop_a* with a specific number of iterations is performed and the PAPR will be compared. If the PAPR is less than the threshold, the signal will be transmitted regardless of the second loop; otherwise the second loop, *Loop_b* with a predefined number of iterations, is executed and the PAPR is similarly calculated. Here, *Loop_a* is repeated 10 times; the reason is that at the end the results should be comparable with the DSI-SLM method, which has been performed with 10 iterations of the dummy insertion loop. The number of *Loop_b* iterations has been determined by trial and error. By increasing the *Loop_b* iteration, the performance is improved, however the processing time is increased too.

When *Loop_a* is performed, a new random dummy is generated and inserted to the signal, but the phase sequence is the same as the last iteration. It should be noted that the number of iterations is specified based on the PAPR reduction requirement and data rate. The value of the PAPR threshold is also based on each standard in wireless broadband. In addition, when *Loop_b* is running, *Loop_a* is repeated. This means that in *Loop_b*, a new random phase sequence will be selected and multiplied to the signal and then a new random dummy is inserted to the signal. When the threshold condition is passed, the signal will be transmitted; however, if the iterations for both loops are performed and the PAPR is still not less than the threshold, the signal with the minimum PAPR among them will be transmitted.

Here, in order to decrease the complexity of the DSI-SLM scheme and increase the spectrum efficiency, a new phase sequence matrix is generated. This new phase sequence is based on the generation of K random values of $\{1 \ -1 \ j \ -j\}$ in which the number of allowed phase factor is 4 and K is the IFFT length or number of subcarriers. It should be noted that if the number of allowed phase factor be specified 2, meaning $\{1 \ -1\}$, then the PAPR performance will be degraded. The phase sequence matrix is given by

$$\hat{B} = \begin{bmatrix} B_{1,1} & ,..., & B_{1,K} \\ \vdots & \vdots & \vdots \\ B_{M,1} & ,..., & B_{M,K} \end{bmatrix}_{[M \times K]} \tag{4.37}$$

where M is the number of iterations in the phase sequence loop, which can be the same as candidate signals. The other forms of the matrix in Equation (4.37) can be shown as follows:

$$\hat{B} = \begin{bmatrix} \overbrace{B_{1,1},...,B_{1,K/Q}}^{Q} & ,..., & \overbrace{B_{1,1},...,B_{1,K/Q}} \\ \vdots & \vdots & \vdots \\ B_{M,1},...,B_{M,K/Q} & ,..., & B_{M,1},...,B_{M,K/Q} \end{bmatrix}_{[M \times K]} \tag{4.38}$$

$$\hat{B} = \begin{bmatrix} \overbrace{B_{1,1},...,B_{1,1}}^{Q}, \overbrace{B_{1,2},...,B_{1,2}}^{Q} ,..., \overbrace{b_{1,K/Q},...,b_{1,K/Q}}^{Q} \\ \vdots & \vdots & \vdots \\ B_{M,1},...,B_{M,1},B_{M,2},...,B_{M,2},...,B_{M,K/Q},...,B_{M,K/Q} \end{bmatrix}_{[M \times K]} \tag{4.39}$$

where Equations (4.38) and (4.39) are interleaved and adjacent phase sequence matrix, respectively, and Q is the division factor.

An interleaved phase sequence is obtained by merging two sequences and an adjacent sequence matrix is created by repeating a smaller matrix. The OPS-DSI scheme uses a random phase sequence matrix as presented in Equation (4.37) due to the highest PAPR performance among them, however, its complexity will be inferior. The main reason that the complexity of the random phase sequence is inferior compared to the adjacent and interleaved phase sequence is due to the symmetric nature of the interleaved and adjacent phase sequences, which requires less memory to store them. In simulation results, the comparison of the PAPR performance will be shown. Since the oversampling factor should be taken into account in order to have the precise PAPR value, the input signal X can be shown as follows:

$$X = [X_0, X_1, ..., X_{KF-1}]_{1 \times KF-1} \tag{4.40}$$

where F is the oversampling factor. To have accurate PAPR, the value of F should be at least 4 (Baxley and Zhou, 2007; Malkin et al., 2008; Wang, Ku et al., 2010).

As oversampling adds zeros to the vector, by multiplying phase sequence B with the input signal X, the only section that counts in the multiplication will be K elements, hence the new phase sequence matrix in Equation (4.37) can be considered and the oversampling factor does not have any effect on these calculations meaning that the oversampling does not affect the complexity and can be ignored.

The next step is to multiply one row of the phase sequence matrix with the input signal. This is performed as follows:

$$\hat{X} = X \otimes B \tag{4.41}$$

where $\otimes$ is the element-by-element product and B is the phase sequence from the Equation (4.37). Each row of the Matrix in Equation (4.37) is multiplied with input signal X to find the optimum phase sequence to minimize the PAPR. The number of iterations that is required to obtain the optimum phase sequence denotes P where P can be any value equal or less than M or in other words $P \leq M$. This condition shows that the optimum phase sequence can be reached at any iteration less than M; however in DSI-SLM or C-SLM, the number of subblocks is fixed to M, which causes a high complexity to the system. Now the dummy sequence is inserted to the product signal. The dummy sequence is selected from a matrix as given by

$$\hat{R} = \begin{bmatrix} R_{1,1} & ,..., & R_{1,L} \\ \vdots & \vdots & \vdots \\ R_{I,1} & ,..., & R_{I,L} \end{bmatrix}_{[I \times L]} \tag{4.42}$$

where L is the dummy sequence length, and I is the number of iterations in the dummy sequence loop from the possible dummy insertion, and R is the random generation of the dummy signal from the values of $\{1\ -1\ j\ -j\}$. The main advantage of inserting dummy signals at this stage is to replace the side information within these signals and also to decrease the number of IFFTs by enhancing the PAPR performance. The dummy signals can increase the PAPR reduction with low complexity. It should be noted that only M number of phase sequences are available to choose from, so $\log_2 M$ number of bits are required to transmit the index of the selected phase sequence similar to DSI-SLM. For example, if $M = 16$, four bits of dummy sequence are indicated to present 1 of 16 indexes of the phase sequence matrix, which has been multiplied with the input signal. Then, the signal is given by

$$S(N) = \left[\hat{X}(K)\right] + \left[R(L)\right] \tag{4.43}$$

The addition in Equation (4.43) means that the insertion of the L dummy signal R to the tail of the input signal X with length K. This is usually performed with multiplexing.

The transmit signal to the IFFT block is given by

$$S_1(n) = \frac{1}{N} \sum_{K=0}^{N-1} S_1(K) \exp(2\pi\ jnk/N) \tag{4.44}$$

In matrix form, Equation (4.44) can be expressed by

$$s = H^{-1}S \tag{4.45}$$

where H is the twiddle factor matrix as follows:

$$H = \frac{1}{N} \begin{bmatrix} 1 & 1 & \cdots & 1 \\ 1 & W^{-1} & \cdots & W^{-(N-1)} \\ \vdots & & & \vdots \\ 1 & W^{-(N-1)} & \cdots & W^{-(N-1)(N-1)} \end{bmatrix}_{N \times N} \qquad (4.46)$$

The PAPR is calculated as follows

$$PAPR = \frac{\max\left[|s(n)|^2\right]}{E\left[|s(n)|^2\right]} \qquad (4.47)$$

Following the PAPR computation, its value will be checked with the threshold in which the threshold value is defined by the IEEE standard. In this case, the PAPR value should be less than 7 dB (Hemphill et al., 2007), which is according to the IEEE standard that is defined for WiMAX applications. By assuming less value for the threshold, then the number of iterations to search for optimum phase sequence and also dummy sequence insertion will be increased, which results in data rate loss due to the long processing that is required to search for the optimum phase sequence. Especially in high data rate applications, the iterations should be taken into consideration.

By performing the first iteration loop for the dummy sequence, if the PAPR is still not less than the threshold, then it goes to the next loop in which the new phase sequence is generated. This operation will be continued until the optimum phase sequence is generated and hence its index i, in matrix B will be transferred together with the input signal as side information. The index i will be located within the dummy signals in order to avoid spectral efficiency degradation.

The derived optimum phase sequence is based on the criteria that the PAPR is less than a threshold or it is minimized. Hence, the criteria of optimum phase sequence can be given by

$$PAPR_{opt} = min \{PAPR[S]\} < PAPR_{th} \qquad (4.48)$$

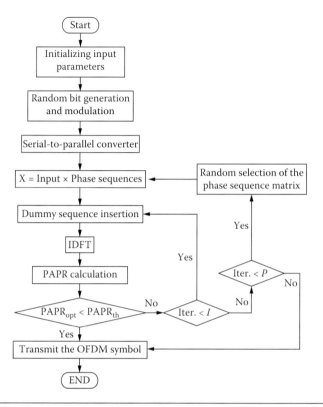

Figure 4.41 Flowchart for the OPS-DSI transmitter.

The flowchart for the OPS-DSI scheme is shown in Figure 4.41. In this flowchart, I is the number of iterations of the DSI loop and P is the number of iterations in the phase sequence loop.

The receiver section of OPS-DSI is designed as shown in Figure 4.42. The received signal is downconverted and passed through the FFT block. From there, the side information and phase sequence is extracted. By the demodulating and decoding processes, the original data stream will be retrieved.

The flowchart for the OPS-DSI receiver is shown in Figure 4.43. By passing the OFDM symbol through the AWGN channel, the time domain signal reaches the low noise amplifier (LNA), which is the first section of the receiver, and following that the signal is downconverted and passed through the analog-to-digital converter. After serial-to-parallel conversion, the FFT is performed and the signal is converted from time domain to frequency domain. The side information will be extracted from the frequency domain at the output of the FFT.

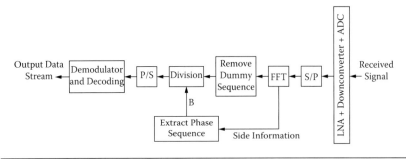

Figure 4.42 Block diagram of the OPS-DSI receiver.

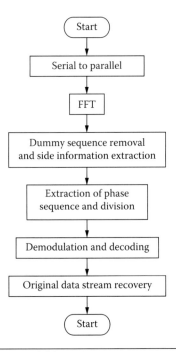

Figure 4.43 Flowchart for the OPS-DSI receiver.

The side information contains the information of the phase sequence that should be applied in order to recover the original signal. Hence, the phase sequence is derived from the side information. Finally, the FFT output signal should be divided to the phase sequence vector to regenerate the input signal.

To verify the performance of the received signal, the BER metric should be measured.

As shown in Figure 4.44, the receiver of the OPS-DSI is almost similar to the DSI-SLM scheme. The difference is in the side

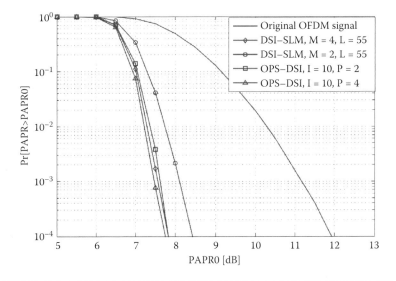

Figure 4.44 CCDF comparison of the OPS-DSI and DSI-SLM schemes ($N = 256$).

information extraction. In OPS-DSI, the side information has to be extracted from the dummy sequence, which contains the phase sequence index. By extracting the address of the phase sequence, the complex division is performed and the original signal is extracted. The extracted signal is demodulated and decoded the same way as the DSI-SLM scheme. Following that, the original data stream is retrieved.

4.7.2 System Performance of the OPS-DSI Scheme

As mentioned earlier about the system performance of DSI-SLM, the TE of DSI-SLM was degraded due to the dummy signal insertion and side information. However, in the proposed OPS-DSI scheme, the TE does not degrade due to replacing the side information bits into dummy signals. If the dummy signals are also replaced with zeros of the OFDM signal, then the TE is still 100%.

4.7.2.1 OPS-DSI Side Information   As mentioned before, the dummy sequence that is used in this scheme includes side information bits. Four bits replaces 55 values of the dummy sequence, which indicates the index of the phase sequence that has been multiplied with the input signal in the transmitter. At the receiver, the side information is accessible by removing the dummy and extracting the last 4 bits from the dummy sequence. Therefore, there is no spectrum efficiency degradation regarding the side information in the OPS-DSI scheme.

4.7.2.2 Advantages and Disadvantages of the Proposed OPS-DSI Scheme The OPS-DSI scheme outperforms DSI-SLM in PAPR performance and complexity. The OPS-DSI scheme has less complexity compared to DSI-SLM because it only needs one IFFT to be implemented. In OPS-DSI, the side information is replaced with the dummy signals and it results in efficiency enhancement compared to DSI-SLM. The other feature of this method is the flexibility to achieve both PAPR performance and have less complexity. For example, by adjusting the number of iterations in the DSI loop and phase sequence loop, different PAPR performance can be achieved. In terms of receiver complexity, both techniques have the same performance. One of the other main advantages of the OPS-DSI scheme is that the iteration number, denoted P, can be equal or less than M, which indicates that the optimum PAPR value can be obtained very fast, which decreases the complexity.

By increasing the number of iterations, the processing time also will be increased, which has an effect in the data rate degradation, which can be considered as a downside of the OPS-DSI scheme.

4.7.2.3 OPS-DSI Computational Complexity The total complexity of the proposed OPS-DSI scheme can be given by

$$T_{OPS-DSI} = 4N\log N + 7N + 2L \qquad (4.49)$$

where N is the IFFT length and L is the dummy sequence length. The total complexity of the C-SLM, DSI-SLM, and OPS-DSI are calculated and Table 4.5 presents the comparison. The CCRR for DSI-SLM and OPS-DSI schemes are calculated based on Equations (4.36) and (4.49), respectively.

Table 4.5 Total Computational Complexity Reduction Comparison

C-SLM	DSI-SLM ($L = 55$)	CCRR$_{DSI-SLM}$	OPS-DSI ($M = 1, L = 55$)	CCRR$_{OPS-DSI}$
10773 ($M = 2$)	10883 ($M = 2$)	−1%	4368	59%
23082 ($M = 4$)	10883 ($M = 2$)	53%	4368	81%
47700 ($M = 8$)	10883 ($M = 2$)	77%	4368	90%
96936 ($M = 16$)	23192 ($M = 4$)	76%	4368	95%

By calculating $T_{\text{OPS-DSI}}$, parameters M and L are kept fixed and are equal to 1 and 55, respectively. The reason is that the OPS-DSI technique with $M = 1$ outperforms the C-SLM method with $M = 16$.

It can be obtained from Table 4.5 that $\text{CCRR}_{\text{OPS-DSI}}$ is 95% when the complexity of the OPS-DSI scheme is compared to the complexity of C-SLM with comparable results. Since the number of IFFTs and complex multiplications in OPS-DSI are reduced compared to DSI-SLM, the CCRR is improved by about 20%.

It should be noted that the number of iterations for both the OPS loop and DSI loop do not have impact on the total complexity. The reason is that although these will affect the processing time which results in data rate loss, from the implementation point of view the complexity will not be affected.

4.7.2.4 Simulation Results and Analysis   In this section, the MATLAB simulation of OPS-DSI scheme is presented. The WiMAX signal based on IEEE 802.16e standard is used to verify the proposed scheme. The m.file codes of the OPS-DSI scheme are available in Appendix B. Similar to DSI and DSI-SLM schemes, based on the IEEE standard, the *PAPR*$_{th}$ in OPS-DSI is set to 7 dB. Simulation results of C-SLM, DSI, DSI-SLM, and OPS-DSI are compared by using the CCDF. There is also a brief study about the BER performance, which requires receiver and channel simulation.

Figure 4.44 shows the PAPR performance comparison between OPS-DSI and DSI-SLM schemes. The length of dummy signal, L, is 55 in both techniques. The number of DSI loop iterations is also the same in both techniques (*DSI iter.* (I) = 10).

The DSI-SLM performance is measured based on the different number of IFFTs, $M = 2$ and 4. As mentioned before, OPS-DSI includes two loops, the DSI loop and OPS loop. In Figure 4.44, two conditions for the OPS loop are considered (*OPS iter.* (P) = 2 and 4).

As shown in Figure 4.44, when the number of OPS iterations (P) is increased from 2 to 4, the PAPR reduction is enhanced by about 0.1 dB. The PAPR performance of OPS-DSI when $S = 10$ and $P = 2$ is almost the same as DSI-SLM when $M = 4$ and $S = 10$.

When the number of subcarriers N is increased from 256 to 1024, the effectiveness of PAPR reduction is degraded due to the more

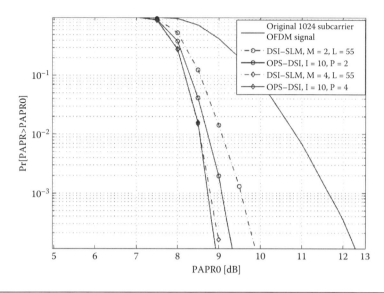

Figure 4.45 CCDF comparison of the OPS-DSI and DSI-SLM schemes ($N = 1024$).

subcarriers that are counted to be in a same symbol, which cause the probability of the number of high peaks to be increased and thus the ratio between peaks and average will be increased too. As shown in Figure 4.45, the original OFDM signal with 1024 subcarriers ($N = 1024$) has PAPR of about 12.5 dB at 10^{-4} CCDF and when DSI-SLM with $M = 2, 4$ is applied, the PAPR is reduced to 9.8 dB and 9 dB, respectively. When the OPS-DSI scheme with $P = 2$ and 4 is applied to the same signal, the PAPR is reduced to about 9.5 dB and 8.9 dBm, respectively.

Figure 4.46 shows the CCDF comparison of the OPS-DSI scheme for different phase sequence formats.

It can be observed that the random phase sequence has the highest PAPR reduction compared to adjacent and interleaved; however, its complexity will be inferior. This is because the interleaved and adjacent phase sequences have alternative phase sequences in their matrices, which results in less complexity.

The OPS-DSI is also compared with the C-SLM technique. Figure 4.47 presents the CCDF result comparison of the OPS-DSI with random phase sequence, $P = 2$ and $I = 10$ and the C-SLM method when $M = 8$ and 16. As shown in Figure 4.47, the original OFDM signal has about 11.8 dB PAPR at 0.01% CCDF. This amount of

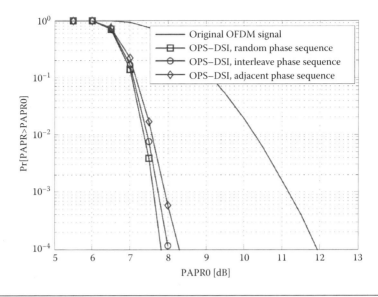

Figure 4.46   PAPR reduction performance of OPS-DSI with different types of phase sequences.

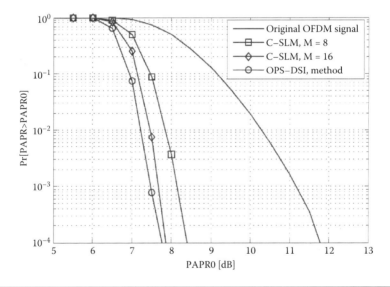

Figure 4.47   PAPR comparison of OPS-DSI with the C-SLM technique.

PAPR is reduced to 8.4 dB and 7.8 dB when the C-SLM technique with $M = 8$ and 16 is applied to the system.

This PAPR reduction performance is improved by applying the proposed OPS-DSI scheme and the PAPR of 7.6 dB can be achieved when $M = 1$.

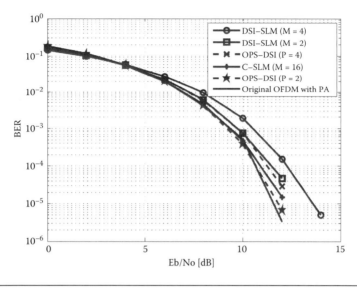

Figure 4.48 BER performance comparison between the OPS-DSI, DSI-SLM, and C-SLM methods.

Therefore, in calculating the total complexity of the OPS-DSI, only one IFFT is considered, which is compared with C-SLM with various values of M.

Figure 4.48 shows a comparison of the BER performance of the C-SLM and the proposed OPS-DSI scheme in AWGN channels. It can be observed that the BER performance of the OPS-DSI scheme is slightly better than DSI-SLM; however, it is insignificant compared to C-SLM.

4.7.2.5 Results Discussion   The CCDF and BER of the proposed OPS-DSI scheme were compared with other methods. It was found that the PAPR performance of OPS-DSI outperforms C-SLM and DSI-SLM, and the complexity is significantly reduced as discussed, which results in less hardware resources and yield system cost reduction.

The reason is that OPS-DSI uses only one IFFT, one complex multiplication, and one complex addition compared to C-SLM, which has 16 numbers of IFFTs and multiplications and DSI-SLM, which requires 4 numbers of IFFTs.

Figure 4.49 shows the time domain comparison between an OFDM signal without and with OPS-DSI, which are indicated with bold and light lines, respectively. As shown in Figure 4.49, by applying OPS-DSI the high peaks (bold) are suppressed, which is shown by light lines.

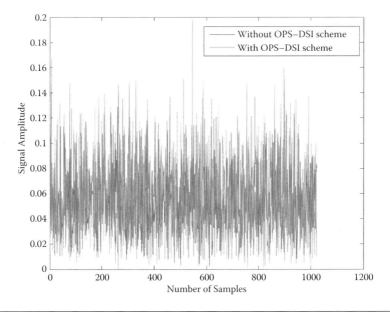

Figure 4.49   The OFDM signal before and after OPS-DSI.

4.8 Summary

The new method of DSI-SLM is proposed in this chapter. According to the algorithm details, DSI-SLM is a combination of modified DSI and C-SLM techniques. The block diagram and the simulation process of DSI-SLM are presented and the results are illustrated.

According to the results discussion, DSI-SLM can reduce PAPR of the OFDM signal by about 3.6 dB when two IFFTs are required for implementation, which leads to a total complexity reduction of 76%. The PAPR performance of DSI-SLM when $M = 2$ is comparable with C-SLM when $M = 8$, which shows the significant reduction in complexity due to fewer IFFTs in the proposed scheme. Low complexity is the main advantage of the proposed DSI-SLM scheme. The complexity of the DSI-SLM is computed and compared with other methods.

The BER degradation is also compared among DSI-SLM, C-SLM, and original OFDM, and according to the results discussion, the BER performance of DSI-SLM is still within the acceptable range of 10^{-4} at $S_b/N_b = 14$ dB.

Since the number of candidate signals or IFFTs is reduced, the number of side information is also reduced, which is another advantage of DSI-SLM. However, transmission of side information has negative effects on bandwidth and data rate. Therefore the side information is still a drawback for DSI-SLM, which should be resolved in order to have high efficient transmission.

Another issue in DSI-SLM is the flexibility, which has potential for improvement. The current design has only one loop. By increasing the number of iterations of this loop, the PAPR performance can be improved. There is a chance to have a more significant result by changing the algorithm to increase the flexibility of the system.

A new method of OPS-DSI is proposed that reduces the PAPR of the OFDM signal as well as the complexity. The algorithm and block diagram of OPS-DSI are explained, and the simulation process is presented. The simulation results are compared with the C-SLM, DSI, and DSI-SLM schemes. The results show that OPS-DSI reduces the PAPR by about 4.2 dB with only 1 IFFT. This reduction is comparable with DSI-SLM with 4 IFFTs and C-SLM with 16 IFFTs.

In the OPS-DSI design, only one IFFT is used, which is the reason that the complexity is reduced even better than the DSI-SLM scheme. The process of generating different candidate signals is performed by repeating the loop for phase sequence generation. Since the loop is repeated until the optimum PAPR is achieved, the phase sequence will be optimum.

Increasing the number of loop iterations has a negative impact on processing delay. This is a drawback that can be overcome by using high-speed processors and techniques like clock partitioning. According to the algorithm, side information bits are inserted within dummy sequence bits and so the problem of bandwidth efficiency degradation due to the side information is resolved.

References

Aburakhia, S. A., Badran, E. F., and Mohamed, D. A. E. 2009. Linear companding transform for the reduction of peak-to-average power ratio of OFDM signals. *IEEE Transactions on Broadcasting* 55(1):155–160.

Armstrong, J. 2002. Peak-to-average power reduction for OFDM by repeated clipping and frequency domain filtering. *Electronics Letters* 38(5):246–247.

Bauml, R. W., Fischer, R. F. H., and Huber, J. B. 1996. Reducing the peak-to-average power ratio of multicarrier modulation by selected mapping. *Electronics Letters* 32(22):2056–2057.

Baxley, R. J., and Zhou, G. T. 2007. Comparing selected mapping and partial transmit sequence for PAR reduction. *IEEE Transactions on Broadcasting* 53(4):797–803.

Cao, R., Jiang, T., and Qin, J. 2007. Study on companding transforms for reduction in PAPR of OFDM signals. *Tien Tzu Hsueh Pao/Acta Electronica Sinica* 35(6):1099–1101.

Chang, L., and Yang, C. 2009. A combined approach of MBAP/PR PAPR reduction and polynomial predistortion for performance enhancement. Proceedings of the 7th International Conference on Information, Communications and Signal Processing (ICICS 2009), Macau, China, December 8–10.

Chang, P., Jeng, S., and Chen, J. (2010). Utilizing a novel root companding transform technique to reduce PAPR in OFDM systems. *International Journal of Communication Systems* 23:447–461.

Chaokuntod, S., Uthansakul, P., and Uthansakul, M. 2009. Fast dummy sequence insertion method for PAPR reduction in WiMAX systems. Proceedings of Asia Pacific of Microwave Conference (APMC 2009), Singapore, December 7–10.

Chen, J. 2010. Application of quantum-inspired evolutionary algorithm to reduce PAPR of an OFDM signal using partial transmit sequences technique. *IEEE Transactions on Broadcasting* 56(1):110–113.

Cooper, S. 2008. Digital radio techniques for energy efficient OFDM base stations, axis network technology (white paper), retrieved March 2009 from http://www.axisnt.com/downloads/DigitalRadioWP.pdf.

Deng, Q., and Zhong, H. 2008. An improved algorithm to reduce PAPR based clipping-and-filtering. Proceedings of the 4th International Conference on Wireless Communications, Networking and Mobile Computing (WiCOM '08), Dalian, China, October 12.

Faulkner, M., and Johansson, M. 1994. Adaptive linearization using predistortion-experimental results. *IEEE Transactions on Vehicular Technology* 43(2):323–332.

Faulkner, M. 2000. An automatic phase adjustment scheme for RF and Cartesian feedback linearizers. *IEEE Transactions on Vehicular Technology* 49(3):956–964.

Foomooljareon, P., and Fernando, W. A. C. 2002. Input sequence envelope scaling in PAPR reduction of OFDM. Proceedings of the 5th International Symposium on Wireless Personal Multimedia Communications, Honolulu, HI, October 27–30.

Foomooljareon, P., Fernando, W. A. C., and Ahmed, K. M. 2003. PAPR reduction of OFDM systems using input sequence envelope scaling. Proceedings of the 57th IEEE Semiannual Conference on Vehicular Technology (VTC 2003-Spring), Jeju, Korea, April 22–25.

Gao, J., Wang, J., and Wang, B. 2009. Peak-to-average power ratio reduction based on cyclic iteration partial transmit sequence. *3rd International Symposium on Intelligent Information Technology Application, IITA 2009* 2:161–164.

Ghassemi, A., and Gulliver, T. A. 2009. Intercarrier interference reduction in OFDM systems using low complexity selective mapping. *IEEE Transactions on Communications* 57(6):1608–1611.

Ghassemi, A., and Gulliver, T. 2010. PAPR reduction of OFDM using PTS and error-correcting code subblocking. *IEEE Transactions on Wireless Communications* 9(3):980–989.

Giannopoulos, T., and Paliouras, V. 2009. A low-complexity PTS-based PAPR reduction technique for OFDM signals without transmission of side information. *Journal of Signal Processing Systems* 56(23):141–153.

Gilabert, P. L., Gadringer, M. E., Montoro, G., Mayer, L., Silveira, D., Bertran, E., and Magerl, G. 2009. An efficient combination of digital predistortion and OFDM clipping for power amplifiers. *International Journal of RF and Microwave Computer-Aided Engineering Archive* 19(5):583–591.

Gregorio, F. H. 2007. Analysis and compensation of nonlinear power amplifier effects in multi-antenna OFDM systems. Helsinki University of Technology, Finland.

Gurung, A. K., AI-Qahtani, F. S., Sadik, A. Z., and Hussain, Z. M. 2008. One-iteration-clipping-filtering (OICF) scheme for PAPR reduction of OFDM signals. Proceedings of International Conference on Advanced Technologies for Communications (ATC 2008), Hanoi, Vietnam, October 6–9.

Han, S. H., and Lee, J. H. 2005. An overview of peak-to-average power ratio reduction techniques for multicarrier transmission. *IEEE Wireless Communications* 12(2):56–65.

Hao, M., and Liaw, C. 2006. A companding technique for PAPR reduction of OFDM systems. Proceedings of International Symposium on Intelligent Signal Processing and Communications (ISPACS '06), Yonago, Japan, December 12–15.

Hao, M., and Liaw, C. 2008. A companding technique for PAPR reduction of OFDM systems. *IEICE Transactions on Communications* E91-B(3):935–938.

Hemphill, E., Summerfield, S., Wang, G., and Hawke, D. 2007. Peak cancellation crest factor reduction reference design. Xilinx Application Note, December 5.

Heo, S., Joo, H., No, J., Lim, D., and Shin, D. 2009. Analysis of PAPR reduction performance of SLM schemes with correlated phase vectors. Proceedings of the IEEE International Symposium on Information Theory (ISIT 2009), Seoul, Korea, June 28–July 3.

Higashinaka, M., Fukui, N., and Kubo, H. 2009. On peak to average power ratio of generalized frequency division multiple access. *IEICE Electronics Express* 6(13):943–948.

Hong, E., and Har, D. 2010. Peak-to-average power ratio reduction in OFDM systems using all-pass filters. *IEEE Transactions on Broadcasting* 56(1):114–119.

Hosseini, I., Omidi, M. J., Kasiri, K., Sadri, A., and Gulak, P. G. 2006. PAPR reduction in OFDM systems using polynomial-based compressing and iterative expanding. Proceedings of IEEE International Conference on Acoustics, Speech and Signal Processing (ICASSP 2006), Toulouse, France, May 14–19.

Hou, J., Ge, J., and Li, J. 2011. Peak-to-average power ratio reduction of OFDM signals using PTS scheme with low computational complexity. *IEEE Transactions on Broadcasting* 57(1):143–148.

Huang, J., Zhou, S., and Willett, P. 2008. Nonbinary LDPC coding for multi-carrier underwater acoustic communications. OCEANS'08 MTS/IEEE Kobe-Techno-Ocean'08–Voyage Toward the Future (OTO'08), Kobe, Japan, April 8–11.

Huang, X., Lu, J., Zheng, J., Letaief, K. B., and Gu, J. 2004. Companding transform for reduction in peak-to-average power ratio of OFDM signals. *IEEE Transactions on Wireless Communications* 3(6):2030–2039.

IEEE 802.16e WiMAX OFDMA Signal Measurements and Troubleshooting. Agilent Application Note 1578, 2011.

IEEE STD 802.16e™-2005. IEEE Standard for local and metropolitan area networks.

Irukulapati, N. V., Chakka, V. K., and Jain, A. 2009. SLM-based PAPR reduction of OFDM signal using new phase sequence. *Electronics Letters* 45(24):1231–1232.

Jayalath, A., and Tellambura, C. 2000. Reducing the peak-to-average power ratio of orthogonal frequency division multiplexing signal through bit or symbol interleaving. *Electronics Letters* 36(13):1161–1163.

Jeon, H. B., No, J. S., and Shin, D. J. 2011. A low-complexity SLM scheme using additive mapping sequences for PAPR reduction of OFDM signals. *IEEE Transactions on Broadcasting* 57(4):866–875.

Jiang, T., Yang, Y., and Song, Y. H. 2005. Exponential companding technique for PAPR reduction in OFDM systems. *IEEE Transactions on Broadcasting* 51(2):244–248.

Jones, A., Wilkinson, T., and Barton, S. 1994. Block coding scheme for reduction of peak to mean envelope power ratio of multicarrier transmission schemes. *Electronics Letters* 30(25):2098–2099.

Jones, D. L. 1999. Peak power reduction in OFDM and DMT via active channel modification. Proceedings of Thirty-Third Asilomar Conference on Signals, Systems, and Computers, Pacific Grove, California, October 24–27.

Kang, S. G., Kim, J. G., and Joo, E. K. 1999. A novel subblock partition scheme for partial transmit sequence OFDM. *IEEE Transactions on Broadcasting* 45(3):333–338.

Katz, A. 2004. Linearizing high power amplifiers. http://www.lintech.com/PDF/hpa.pdf.

Kim, J., Han, S., and Shin, Y. 2008. A robust companding scheme against nonlinear distortion of high power amplifiers in OFDM systems. Proceedings of IEEE 67th Vehicular Technology Conference-Spring (VTC 2008), Marina Bay, Singapore, May 11–14.

Kim, J., and Shin, Y. 2008. An effective clipped companding scheme for PAPR reduction of OFDM signals. Proceedings of the IEEE International Conference on Communications (ICC 2008), Beijing, China, May 19–23.

Kim, S. W., Chung, J. K., and Ryu, H. G. 2006. PAPR reduction of the OFDM Signal by the SLM-based WHT and DSI method. IEEE Region 10 Conference (TENCON-2006), Hong Kong, China, November 14–17.

Krongold, B. S., and Jones, D. L. 2003. PAR reduction in OFDM via active constellation extension. *IEEE Transactions on Broadcasting* 49(3):258–268.

Kumar, N. S. L. P., Banerjee, A., and Sircar, P. 2007. Modified exponential companding for PAPR reduction of OFDM signals, Proceedings of IEEE Wireless Communications and Networking Conference (WCNC 2007), Hong Kong, March 11–15.

Kwon, J. W., Park, S. K., and Kim Y. 2009. Peak-to-average power ratio reduction by the partial shift sequence method for space-frequency block coded OFDM systems. Proceedings of 2009 IEEE International Conference on Network Infrastructure and Digital Content (IEEE IC-NIDC2009), Beijing, China, November 6–8.

Kwon, U., Kim, D., and Im, G. 2009. Amplitude clipping and iterative reconstruction of MIMO-OFDM signals with optimum equalization. *IEEE Transactions on Wireless Communication* 8(1):268–277.

Le Goff, S. Y., Al-Samahi, S. S., Khoo, B. K., Tsimenidis, C. C., and Sharif, B. S. 2009. Selected mapping without side information for PAPR reduction in OFDM. *IEEE Transactions on Wireless Communication* 8(7):3320–3325.

Le Goff, S. Y., Khoo, B. K., Tsimenidis, C. C., and Sharif, B. S. 2008. A novel selected mapping technique for PAPR reduction in OFDM systems. *IEEE Transactions on Communication* 56(11):1775–1779.

Lee, B. M., and de Figueiredo, R. J. P. 2010. MIMO-OFDM PAPR reduction by selected mapping using side information power allocation. *Digital Signal Processing* 20(2):462–471.

Leung, S. H., Ju, S. M., and Bi, G. G. 2002. Algorithm for repeated clipping and filtering in peak-to-average power reduction for OFDM. *Electronics Letters* 38(25):1726–1727.

Li, Ch. P., Wang, S. H., and Wang, Ch. L. 2010. Novel low-complexity SLM schemes for PAPR reduction in OFDM systems. *IEEE Transactions on Signal Processing* 58(5):2916–2921.

Lim, D. W., No, J. S., Lim, C. W., and Chung, H. 2005. A new SLM OFDM scheme with low complexity for PAPR reduction. *IEEE Signal Processing Letters* 12(2): 93–96.

Luo, J., Keusgen, W., and Kortke, A. 2008. Optimization of time domain windowing and guardband size for cellular OFDM systems. Proceedings of the 68th Semi-Annual IEEE Vehicular Technology (VTC 2008-Fall), Alberta, Canada, September 21–24.

Malkin, M., Krongold, B., and Cioffi, J. M. 2008. Optimal constellation distortion for PAR reduction in OFDM systems. Proceedings of IEEE 19th International Symposium on Personal, Indoor and Mobile Radio Communications (PIMRC 2008). Cannes, France, September 15–18.

May, T., and Rohling, H. 1998. Reducing the peak-to-average power ratio in OFDM radio transmission systems, Proceedings of 48th IEEE Vehicular Technology Conference (VTC 1998), Ottawa, Canada, May 18–21.

Muller, S. H., and Huber J. B. 1997a. A comparison of peak power reduction schemes for OFDM. IEEE Global Telecommunications Conference (GLOBECOM '97), Erlangen, Germany, November 3–8.

Muller, S. H., and Huber, J. B. 1997b. A novel peak power reduction scheme for OFDM. Proceedings of 8th IEEE International Symposium on Personal, Indoor and Mobile Radio Communications (PIMRC '97), Helsinki, Finland, September 1–4.

Muller, S. H., and Huber, J. B. 1997c. OFDM with reduced peak-to-average power ratio by optimum combination of partial transmit sequences. *Electronics Letters* 33(5):368–369.

Naeiny, M. F., and Marvasti, F. 2011. Selected mapping algorithm for PAPR reduction of space-frequency coded OFDM systems without side information. *IEEE Transactions on Vehicular Technology* 60(3):1211–1216.

Nikookar, H., and Lidsheim, K. S. 2002. Random phase updating algorithm for OFDM transmission with low PAPR. *IEEE Transactions on Broadcasting* 48(2):123–128.

Ochiai, H. 2003. Performance analysis of peak power and band-limited OFDM system with linear scaling. *IEEE Transactions on Wireless Communications* 2(5):1055–1065.

Ochiai, H., and Imai, H. 1997. Block coding scheme based on complementary sequences for multicarrier signals. *IEICE Transactions on Fundamentals of Electronics, Communications and Computer Sciences* 80(11):2136–2143.

Ochiai, H., and Imai, H. 2002. Performance analysis of deliberately clipped OFDM signals. *IEEE Transactions on Communications* 50(1):89–101.

Pratt, T. G., Jones, N., Smee, L., and Torrey, M. 2006. OFDM link performance with companding for PAPR reduction in the presence of nonlinear amplification. *IEEE Transactions on Broadcasting* 52(2):261–267.

Qian, H., Xiao, Ch., Chen, N., and Zhou, G. T. 2005. Dynamic selected mapping for OFDM. Proceedings of IEEE International Conference on Acoustics, Speech, and Signal Processing (ICASSP '05). Philadelphia, March 18–23.

Raab, F. H., Asbeck, P., Cripps, S., Kenington, P. B., Popovic, Z. B., Pothecary, N., Sevic, J. F., and Sokal, N. O. 2002. Power amplifiers and transmitters for RF and microwave. *IEEE Transactions on Microwave Theory and Techniques* 50(3):814–826.

Rostamzadeh, M., Vakily, V. T., and Moshfegh, M. 2008. PAPR reduction in WPDM and OFDM systems using an adaptive threshold companding scheme. Proceedings of the 5th International Multi-Conference on Systems, Signals and Devices (SSD '08), Amman, Jordan, July 20–23.

Ryu, H. G., Lee, J. E., and Park, J. S. 2004. Dummy sequence insertion (DSI) for PAPR reduction in the OFDM communication system. *IEEE Transactions on Consumer Electronics* 50(1):89–94.

Schniter, P. 2004. Low-complexity equalization of OFDM in doubly selective channels. *IEEE Transactions on Signal Processing* 52(4):1002–1011.

Sharma, P. K., and Basu, A. 2010. Performance analysis of peak-to-average power ratio reduction techniques for wireless communication using OFDM signals. Proceedings of the International Conference on Advances in Recent Technologies in Communication and Computing (ARTCom), India, October 16–17.

Sulaiman, S. A., Badran, E. F., and Mohamed, D. A. E. 2007. A comparison between clipping and μ-law companding schemes for the reduction of peak-to-average power ratio of OFDM. Proceedings of 24th National Radio Science Conference (NRSC 2007), Cairo, Egypt, March 13–15.

Takyu, O., Ohtsuki, T., and Nakagawa, M. 2006. Companding system based on time clustering for reducing peak power of OFDM symbol in wireless communications. *IEICE Transactions on Fundamentals of Electronics, Communications and Computer Sciences* E89-A(7):1884–1891.

Tan, C. E., and Wassell, I. J. 2003. Data bearing peak reduction carriers for OFDM systems. Information, Proceedings of the 2003 Joint Conference of the 4th International Conference on Communications and Signal Processing the Fourth Pacific Rim Conference on Multimedia, Singapore, December 15–18.

Tang, L., Li, L., and Zhang, Q. 2009. An improved SLM method based on chaotic phase sequences. Proceedings of 5th International Conference on Wireless Communications, Networking and Mobile Computing (WiCOM 2009), Beijing, China, September 24–26.

Tao, J., and Li, P. 2010. An improved SLM method for PAPR reduction in OFDM systems. Proceedings of IET 3rd International Conference on Wireless, Mobile and Multimedia Networks (ICWMNN 2010), China, September 26–29.

Tellado, J. 2000a. *Multicarrier Modulation with Low PAR: Applications to DSL and Wireless.* Dordrecht, Netherlands: Kluwer Academic.

Tellado, J. 2000b. Peak to average power reduction for multicarrier modulation. Doctoral dissertation, Stanford University, California.

Urban, J., and Marsalek, R. 2008. OFDM PAPR reduction by partial transmit sequences and simplified clipping with bounded distortion. Proceedings of 18th International Conference on Radioelektronika, Prague, Czech Republic, April 24–25.

Uthansakul, P., Chaokuntod, S., and Uthansakul, M. 2009. Fast dummy sequence insertion method for PAPR eeduction in WiMAX systems. *International Journal of Electronics, Communications and Computer Engineering* 1:3.

Van Nee, R., and De Wild, A. 1998. Reducing the peak-to-average power ratio of OFDM. Proceedings of 48th IEEE Vehicular Technology Conference (VTC 1998), Ottawa, Canada, May 18–21.

Varahram, P., Al-Azzo, W. F., and Ali, B. M. 2010. A low complexity partial transmit sequence scheme by use of dummy signals for PAPR reduction in OFDM systems. *IEEE Transactions on Consumer Electronics* 56(4):2416–2420.

Vijayarangan, V., and Sukanesh, D. 2009. An overview of techniques for reducing peak to average power ratio and its selection criteria for orthogonal frequency division multiplexing radio systems. *Journal of Theoretical and Applied Information Technology* 5(1):25–36.

Wang, Ch. L., and Ku, S. J. 2006. A low-complexity companding transform for peak-to-average power ratio reduction in OFDM systems. Proceedings of IEEE International Conference on Acoustics, Speech and Signal Processing (ICASSP 2006), Toulouse, France, May 14–19.

Wang, Ch. L., Ku, Sh. J., and Yang, Ch. J. 2010. A low-complexity PAPR estimation scheme for OFDM signals and its application to SLM-based PAPR reduction. *IEEE Journal of Selected Topics in Signal Processing* 4(3):637–645.

Wang, J., Guo, Y., and Zhou, X. 2009. PTS-clipping method to reduce the PAPR in ROF-OFDM system. *IEEE Transactions on Consumer Electronics* 55(2):356–359.

Wang, J., Wu X., Mao, Z., and Zhou, B. 2009. Phase factor combination scheme based on pipeline in time domain IP-PTS for PAPR reduction of OFDM systems. Proceedings of 15th Asia-Pacific Conference on Communications (APCC 2009), Shanghai, China, October 8–10.

Wang, J. S., Lee, J. H., Park, J. Ch., Song, I., and Kim, Y. H. 2010. Combining of cyclically delayed signals: A low-complexity scheme for PAPR reduction in OFDM systems. *IEEE Transactions on Broadcasting* 56(4):577–583.

Wang, K., Hao, J., Wang, L., and Li, H. 2009. Phase factor sequences algorithm in partial transmit sequence. *Transactions of Tianjin University* 15(1):23–26.

Wang, L., Kang, Y., and Bin, X. 2009. Using the union algorithm of SLM and PTS to reduce PAPR in OFDM system. Proceedings of Second ISECS International Colloquium on Computing, Communication, Control, and Management (CCCM 2009), Sanya, China, August 8–9.

Wang, L., and Liu, J. 2011. PAPR reduction of OFDM signals by PTS with grouping and recursive phase weighting methods. *IEEE Transactions on Broadcasting* 57(2):299–306.

Wang, S. H., Sie, J. Ch., Li, Ch. P., and Chen, Y. F. 2011. A low-complexity PAPR reduction scheme for OFDMA uplink systems. *IEEE Transactions on Wireless Communications* 10(4):1242–1251.

Wang, X., Tjhung, T. T., and Ng, C. S. 1999. Reduction of peak-to-average power ratio of OFDM system using a companding technique. *IEEE Transactions on Broadcasting* 45(3):303–307.

Wei, G., Hu, L., and Yu, H. 2006. PAR reduction for OFDM and DMT signals based on distribution transform. *Chinese Journal of Electronics* 15(2):282–286.

Wilkinson, T. A., and Jones, A. E. 1995. Minimization of the peak to mean envelope power ratio of multicarrier transmission schemes by block coding. Proceedings of the IEEE Vehicular Technology Conference, Chicago, Illinois, July 25–28.

Wu, X., Wang, J., Mao, Z., and Zhang, J. 2010. Conjugate interleaved partitioning PTS scheme for PAPR reduction of OFDM signals. *Circuits, Systems, and Signal Processing* 29(3):499–514.

Wu, X., Zhou, B., Wang, J., and Mao, Z. 2010. Low complexity time domain interleaved partitioning partial transmit sequence scheme for peak-to-average power ratio reduction of orthogonal frequency division multiplexing systems. *Wireless Personal Communications* 53(3).

Xue, K., Yang, H., and Su, Sh. 2009. The clipping noise and PAPR in the OFDM system. Proceedings of the International Conference on Communications and Mobile Computing (CMC '09), Kunming, Yunnan, China, January 6–8.

Yan, B., Zhang, H., Yang, Y., Hu, Q., and Qiu, M. 2009. An improved algorithm for peak-to-average power ratio reduction in MIMO-OFDM systems. 2009 International Conference on Wireless Communications and Signal Processing (WCSP).

Yang, L., Li, S. Q., Soo, K. K., and Siu, Y. M. 2009. SLM with non-unit magnitude phase factors for PAPR reduction in OFDM. Proceedings of International Conference on Communications, Circuits and Systems (ICCCAS 2009), San Jose, CA, July 23–25.

Yang, L., Soo, K., Siu, Y., and Li, S. 2008. A low complexity selected mapping scheme by use of time domain sequence superposition technique for PAPR reduction in OFDM system. *IEEE Transactions on Broadcasting* 54(4):821–824.

Youngoo, Y., and Bumman, K. 1999. A new linear amplifier using low-frequency second-order intermodulation component feedforwarding. *IEEE Microwave and Guided Wave Letters* 9(10):419–421.

Zhang, L., and Thibault, L. 2010. Performance evaluation of mobile DAB receivers in enhanced packet mode. *IEEE Transactions on Consumer Electronics* 56(4):2115–2122.

Zhou, Y., and Jiang, T. 2009. A novel clipping integrated into ACE for PAPR reduction in OFDM systems. International Conference on Wireless Communications and Signal Processing (WCSP 2009), Nanjing, China, November 13–15.

5

PEAK-TO-AVERAGE POWER
RATIO IMPLEMENTATION

5.1 Introduction

One of the main factors to choose the peak-to-average power ratio (PAPR) technique is by verifying the feasibility of the PAPR technique and to ensure that the technique can be implemented in a test bed platform. There are few works that focus on the implementation of the PAPR technique. Most of the PAPR techniques that have been proposed until now suffer from feasibility. The main reason is the complexity of those techniques are mainly high. Today's field programmable gate array (FPGA) technology has limited hardware resources. The main blocks in FPGAs are random access memory (RAM), multipliers, first input first output (FIFO), and adders. The most recent FPGAs are based on a system on chip (SOC) in which it has an Acorn RISC Machine (ARM) processor to integrate the C and VHSIC Hardware Description Language (VHDL). This is important in the sense that some PAPR techniques involved with computation within the iteration loop come from the feedback to optimize the PAPR value. Hence, in implemenation the processing needs to be done in an ARM processor. The conventional FPGAs had to microblaze or power the PC processor where the speed and size were less than the ARM processor. In the view of implementation where the hardware resource is the main concern and directly affects the system cost, the PAPR complexity is one of the main parameters used to evaluate the performance of it. But at the same time two other metrics, PAPR performance and bit error rate (BER), need to be considered. There is always a trade-off between these parameters. Some PAPR techniques enhance the PAPR performance, however, the complexity issue has not been considered. Some techniques have also not considered the complexity with respect to hardware resources. This is due to the fact that there is a distinction between computational complexity and hardware complexity. Thus, we have defined two metrics in order to evaluate the complexity arising from the PAPR techniques: computational complexity reduction ratio (CCRR) and hardware complexity reduction ratio (HCRR). There is a major difference between these parameters. CCRR is mainly the complexity of the calculation in PAPR together with the number of multiplications and additions involved, especially in inverse fast Fourier transform (IFFT). However, the HCRR is a different value due to the fact that the number of multipliers in the feedback loop in the iteration only accounts for one time rather than the CCRR in which multipliers have to be counted for each iteration.

5.2 Software Implementation Design

In Figure 5.1, it is observed that the software implementation block is divided into two main sections: MATLAB simulation and C++ implementation. The two layers of implementation relate closely to each other; the MATLAB simulation requires the C++ implementation level to stay closer to the digital signal processor (DSP) hardware for hardware implementation. On the other hand, C++ implementation needs the MATLAB simulation layer to check and verify its

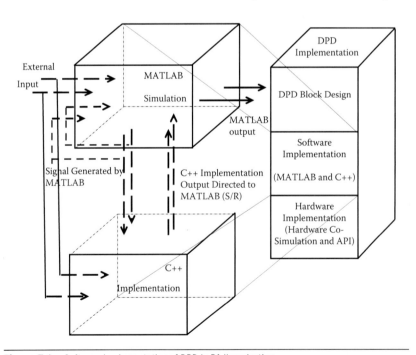

Figure 5.1 Software implementation of DPD in PA linearization.

output, thus plotting output graphs for the user to understand and check the results of the digital predistortion (DPD).

5.2.1 MATLAB Simulation Design

The data type involved in the MATLAB simulation block should all be in type double. Input and outputs should be considered to have 1024 memory slots, each is double. The initial stage of the implementation is tested by using the signal data randomly generated within MATLAB.

5.2.2 C++ Implementation Design

The C++ implementation block first uses the random signal generated by MATLAB to test its workability. External input from real sample signal data of different types (e.g., wideband code division multiple access [WCDMA]) are then fed in to be digitally predistorted. The output is then directed to MATLAB for further graph plotting to check the efficiency of the system.

There are basically four classes created in C++ implementation: AdaptationBlock, SalehModelAmp, ComplexMultiplier, and IQSR. Each class has its own class members, which have specific functionalities that provide code sustainability and maintainability both to the user and the developer for future works. The syntax and naming follows the Google C++ naming convention to avoid confusion in class member naming and also file naming.

5.2.3 Implementation Platform

The hardware implementation design involves the participation of the DSP in the DPD PA (power amplifier) linearization method. The purpose of involving the usage of DSP is to test the workability of the algorithm design in a near realistic environment. In this project, the Microblaze soft processor developed by Xilinx is used due to its flexibility to be formed and built according to the user or developer's design requirements. Therefore, there are no hardware resources wasted for the final implementation of the design.

Figure 5.2 shows the components of the hardware implementation block. The XPS and SDK forms the Xilinx EDK used for programming the DSP. The previously developed C++ implementation design can be

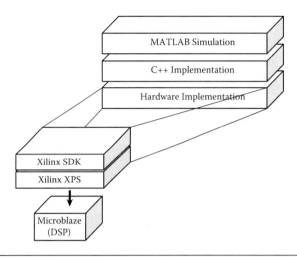

Figure 5.2 Hardware implementation components.

used to directly communicate with the created Microblaze processor by creating a project in the SDK environment, which contains the C++ coding developed in the C++ implementation development stage.

5.3 Hardware Complexity

In this section, the hardware complexity analysis of the proposed scheme is investigated. It should be noted that the term complexity in this chapter means the hardware resource consumption, which is different from the computational complexity in Baxley and Zhou (2007).

The total hardware complexity of the conventional partial transmit sequence (C-PTS) considering the oversampling factor $S = 1$, can be expressed as follows:

$$T_{C-PTS} = 3VN/2 \log N + 3VN/2 \tag{5.1}$$

The complexity in Equation (5.1) accounts for the total complex addition and multiplication; however, a complex multiplication consists of four real multiplications and two real additions, also a complex addition requires two real additions.

Whereas for the enhanced PTS, the total complexity is given by

$$T_{EPTS} = 3/4VN \log N + 3VN/2 \tag{5.2}$$

where V is the number of subblocks.

It is clear that Equation (5.1) and Equation (5.2) are comprised of two parts: the first section is the complexity in respect to the IFFT itself and the second part is related to the searching algorithm complexity. Most of the research has not considered the second part, which results in a miscalculation of the complexity.

The complexity is considered for only IFFT (Ryu et al., 2004), which is not the real scenario. Hence in this chapter, the total complexity is applied in our calculation to have a fair comparison. The total complexity of the dummy signal insertion (DSI)–PTS method is given by

$$T_{DSI-PTS} = 3/4VN \log N + 3VN/2 + L \qquad (5.3)$$

The total complexity of the proposed OPSM scheme can be calculated by

$$T_{OPSM} = 3/4VN \log N + 3VN/2 + L \qquad (5.4)$$

It should be noted that the number of iterations in the DSI loop can be ignored as it only affects the data rate in which the higher number of iterations cause more data rate degradation.

This is due to the higher computational delay in order to find the minimum PAPR; however, the complexity is the hardware resource consumption, which only considers the number of hardware resources.

Here, a new metric called the hardware complexity reduction ratio (HCRR) is defined as follows:

$$HCRR = \left(1 - \frac{Complexity\ of\ OPSM\ scheme}{Complexity\ of\ C - PTS}\right) \times 100\% \qquad (5.5)$$

Table 5.1 gives the hardware complexity values for the C-PTS and the proposed OPSM scheme when $N = 512$ and $W = 2$. It can be observed that the total complexity of DSI-PTS and the optimum phase sequence matrix (OPSM) scheme is the same; however, in simulation results it will be shown that OPSM outperforms DSI-PTS in terms of PAPR performance.

Table 5.1 Hardware Complexity of the Proposed OPSM Scheme and the Conventional PTS When $N = 512$ and $L = 55$

	NUMBER OF SUBBLOCKS	C-PTS	OPSM	HCRR
Total Complexity	$V = 4$	30775	16951	44.9%
	$V = 8$	61495	33847	44.9%

The other important parameter that needs to be considered when calculating the complexity of the proposed scheme is the dummy sequence insertion; however, the insertion of the dummy sequence does not deter the IFFT length. Hence, according to Equation (5.5), it can be concluded that the number of IFFTs is halved. From the simulation results, the PAPR performance of the proposed scheme with the number of IFFT half of the C-PTS is almost the same as the C-PTS. This is examined for different numbers of subblocks and proves that applying the proposed scheme makes the number of IFFT to be halved compared to the C-PTS because it leads to the same PAPR performance.

5.4 Hardware Implementation

The main part of implementing the DSI-EPTS scheme in FPGA is the IFFT block, which is expressed in discrete form in Equation (5.7). The algorithm chosen for implementing IFFT is Radix-2, because it is smaller in size than the pipeline method but has a longer processing time. IFFT is the reverse of FFT and is computed by a phase factor conjugation of FFT. The processing of the I/O signals is not simultaneous, thus the data is first loaded to FFT and stored in RAMs. During the calculation process, a new signal cannot be loaded. The twiddle factors are fixed in Read-Only Memory (ROM), which will be used later to create the IFFT function.

The procedure to implement the DSI-EPTS can be described as follows. First, the input signal, which is a continuous sample, is generated and saved in the memory. The memory unit can be random access memory (RAM), which transfers the data sequence by using the FIFO block. Because in the DSI-EPTS scheme the phase sequence shall be multiplied with the subblocks, hence input samples are replicated in order to multiply with phase sequence and depending on the number of iterations, the input samples are continuously repeated. It should be noted that the input samples and phase sequence are generated and multiplied to execute the EPTS scheme.

The phase sequences are reshaped to a one-dimensional vector in order to perform the searching operation. The output of the IFFT experiences some delay due to the nature of the implementation process in which each of the 256 samples (orthogonal frequency division multiplexing [OFDM] symbol) has to be stored in the memory until all the real and imaginary samples are captured.

This process causes 1291 samples to delay. This delay depends on the type of IFFT implementation whether it is Radix 2 or Radix 4, Streaming IO, or single. Therefore, the delay should be compensated in order to be synchronized with the phase sequence. The output of the IFFT block is followed with the complex multiplier, which consists of four real multipliers, one addition, and one subtraction. The output result can be captured and saved in the memory. Finally, the PAPR is calculated for each OFDM symbol. The PAPR calculation is considered in the hardware resource, but here the PAPR is calculated offline and is not considered in the hardware resource performance. The minimum PAPR value has to be transmitted, which is based on the criteria expressed in Equation (5.25). In order to retrieve the original signal at the receiver, side information has to be transmitted with the signal, which causes transmission efficiency degradation.

Table 5.2 shows the hardware resource of the C-PTS method. It can be observed that the hardware resource consumption is quite high and by increasing the number of subblocks (V) to obtain the better PAPR reduction, the hardware resource increases dramatically.

Table 5.3 shows the hardware resource consumption of the DSI-EPTS scheme in FPGA. The hardware resource consumption of the proposed DSI-EPTS scheme is less than C-PTS, which makes

TABLE 5.2 Hardware Resource Estimation of C-PTS
($N = 512$, $V = 4$, $W = 2$)

HARDWARE	RESOURCES USED	PERFORMANCE (%)
Slices	10494	68
DSP48 slices	80	41
LUT-FF	17091	55
IO Utilization units	168	37

TABLE 5.3 Hardware Resource Estimation of DSI-EPTS
($N = 512$, $V = 2$, $W = 2$)

HARDWARE	RESOURCES USED	PERFORMANCE (%)
Slices	8097	26
DSP48 slices	73	38
LUT-FF	12477	40
IO Utilization units	203	45

the cost of the system drop. The DSP48 slices consume most of the hardware resource allocation. In fact, the number of multipliers is high due to the high number of multiplication processes in the DSI-EPTS scheme. Compared with the results of Table 5.2, it can be observed that the hardware resource performance of the DSI-EPTS scheme is better than the C-PTS due to the smaller number of IFFT operations.

Tables 5.4 and 5.5 present the power consumption report for the C-PTS and DSI-EPTS, respectively. These tables are generated from the hardware resource consumption and computational process that results in a different current that is derived during the implementation process. From Table 5.5, the DSI-EPTS scheme consumes a total power of 1.17006 watts, whereas the C-PTS power consumption is 1.4197 according to Table 5.4.

The data path delay of the C-PTS ($V = 4$) and DSI-EPTS ($V = 2$) are 14.472 ns and 15.865 ns, respectively. It can be concluded that DSI-EPTS consumes less power due to the smaller number of IFFTs

TABLE 5.4 Power Consumption Report of the C-PTS Scheme (Ambient Temperature = 33°C)

NAME	C-PTS ($V = 4$); POWER CONSUMPTION (W)
Logic	0.01123
Signals	0.0000
DSP	0.00525
Total quiescent power	0.63124
Total dynamic power	0.78848
Total power	1.41972

Table 5.5 Power Consumption Report of the DSI-EPTS Scheme (Ambient Temperature = 33°C)

NAME	DSI-EPTS ($V = 2$); POWER CONSUMPTION (W)
Logic	0.01084
Signals	0.02905
DSP	0.0045
Total quiescent power	0.62078
Total dynamic power	0.54928
Total power	1.17006

with a slightly higher processing time, which is due to the higher number of phase sequences in the phase sequence matrix.

5.5 Field Programmable Gate Array

In this section, the field programmable gate array (FPGA) prototype of the two proposed schemes, DSI-SLM (dummy signal insertion with selected mapping) and OPS-DSI (optimum phase sequence with dummy sequence insertion), are presented. First, the algorithm is modeled by using Simulink tools in the MATLAB environment. In order to generate FPGA prototypes of these models, a tool called the Xilinx System Generator is used to generate direct Very High Speed Integrated Circuit (VHSIC) Hardware Description Language (VHDL) by means of the Xilinx Blocksets (Shanblatt and Foulds, 2005). The System Generator and Xilinx blocksets are specialized toolkits for Xilinx FPGAs. Second, the DSI-SLM scheme prototype is designed in order to verify the simulation results. To successfully implement these schemes, both the conventional selective mapping (C-SLM) method and DSI method are implemented individually by using the same hardware platform. To prototype these schemes, the main block of the OFDM system, which is the IFFT, is implemented and its performance is verified separately. Finally, the implementation results of the proposed schemes are compared with the simulation results to verify the hardware prototype. The hardware resource consumption is calculated for both schemes and compared with the C-SLM method. Moreover, the receiver section of the proposed schemes is also implemented by using the Xilinx tools to verify the BER performance. To prototype the receiver, the complex divider is designed and implemented separately. The designed complex divider outperforms the conventional complex divider technique in terms of accuracy and the hardware resource.

FPGAs are configurable and reprogrammable digital logic devices, and programming code is usually written in HDL.

The Virtex-5 FXT Evaluation Kit is used for implementation and has the Xilinx Virtex-5 XC5VFX30T-FF665 FPGA chip. This board also has 64 MB DDR2 SDRAM memory and 16 MB FLASH memory with a variety of I/O gates, which makes it suitable for a typical implementation.

5.5.1 *The System Generator Tool*

System Generator is a Xilinx tool that enables the use of the MathWorks Simulink environment for FPGA design. A set of Xilinx specific blocksets are used to capture the design in Simulink and at the same time automatically translates designs into hardware implementations. This translation can be done by the hardware Co-Simulation tool. This simulation tool captures the design in Simulink and automatically run it on the FPGA development board. Over 20 hardware platforms are supported by this technique, including Xilinx Virtex-5, which is used in this project.

5.5.2 *System Generator Design Flow*

An executable algorithm should be developed by using the standard Xilinx blocksets that are accessible from the Simulink library. The System Generator can be used to specify the hardware target and produce a bit stream for programming the FPGA. The flowchart in Figure 5.3 shows the prototype flow. The simulation is performed in MATLAB. The implementation process is carried out by using MATLAB Simulink, AccelDSP, Integrated Software Environment (ISE), and System Generator tools. The components of the Xilinx library can be combined with other Simulink blocks, but only those subsystems denoted as Xilinx blocks and Xilinx Blockset

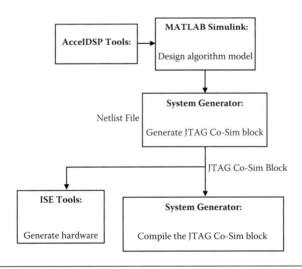

Figure 5.3 Prototype flow diagram.

can be translated by System Generator into a hardware realization. The generation process is controlled from the System Generator Graphical User Interface (GUI), which allows the user to choose the target FPGA device and other implementation options. The System Generator translates the Simulink model into an IP library module and converts it to a hierarchical VHDL netlist. This VHDL netlist file can be translated to hardware realization by using ISE tools.

5.6 The Prototype of the Dummy Signal Insertion with Selected Mapping Scheme

As described earlier, Virtex-5, XC5vfx30t-1ff665 FPGA board is chosen as a typical FPGA evaluation kit for a hardware test. The main block of the proposed dummy signal insertion with selected mapping (DSI-SLM) scheme is the IFFT. Hence, the implementation of the IFFT is explained first. Then, other features of the DSI-SLM are discussed. The second proposed method, OPS-DSI, is also implemented and the results are compared with the simulation. Moreover, the receiver is designed for DSI-SLM and OPS-DSI schemes and the BER performance is verified. To design the receiver, the implementation of the FFT block is crucial, that is, the inverse of the IFFT.

5.6.1 The Inverse Fast Fourier Transform Prototype

This section introduces the fundamentals of the inverse fast Fourier transform (IFFT) block prototype at the transmitter and the FFT block at the receiver. The basic equations for the FFT and the IFFT are given by

$$X(k) = \sum_{n=1}^{N-1} x(n)e^{-j2\pi kn/N}, k = 0, ..., N-1 \qquad (5.6)$$

$$x(n) = \frac{1}{N} \sum_{n=1}^{N-1} X(k)e^{-j2\pi kn/N}, n = 0, ..., N-1 \qquad (5.7)$$

where N is the transform size or the number of sample points in the data frame and $j = \sqrt{-1}$. $X(k)$ is the frequency output of the FFT at kth point where $k = 0, 1, \ldots, N-1$ and $x(n)$ is the time sample at nth point with $n = 0, 1, \ldots, N-1$. Due to the symmetry of the exponential matrix $e^{-j2\pi kn/N}$ it can be represented as the twiddle factor, which is shown by W_N^{nk}. The computation can be performed faster by using a twiddle factor because it depends on the number of points used and there is no need to recalculate, and the values can be referred to as a matrix of twiddle factors. Because the transform time is crucial to the FFT process, there is always a trade-off between the core size and the transform time. In Xilinx there are four architectures: Pipelined-Streaming I/O, Radix4-Burst I/O, Radix2-Burst I/O, and Radix2-Lite-Burst I/O. They each have different features to cover different time and size requirements.

In Pipelined Streaming I/O architecture, the data is continuously processed. Radix4 uses an iterative approach to process the data, which is loaded and processed separately. It is smaller in size than the pipelined solution but has a longer transform time. The third architecture has the same iterative approach as Radix4 but has a longer transform time. Radix2 is based on decimation in frequency (DIF) and separates the input data into two halves of

$$X(0), X(1), \ldots, X\left(\frac{N}{2}-1\right) \text{ and } X\left(\frac{N}{2}\right), X\left(\frac{N}{2}+1\right), \ldots, X(N-1)$$

$$(5.8)$$

The FFT formula for both even and odd conditions can be written in two summations as follows:

$$X(k) = \sum_{n=0}^{\frac{N}{2}-1} a(n)W_N^{nk}, \text{ where } a(n) = x(n) + x\left(n + \frac{N}{2}\right)$$

and $$(5.9)$$

$$X(2k+1) = \sum_{n=0}^{\frac{N}{2}-1} b(n)W_N^{nk}, \text{ where } b(n) = x(n) - x\left(n + \frac{N}{2}\right)$$

where $Y(0) = X(0) + X(1)$ and $Y(0) = X(0) - X(1)$, respectively. This FFT flow graph is called a butterfly graph. When the number of points is increased, the butterfly is expanded. The Radix2 scheme, the third

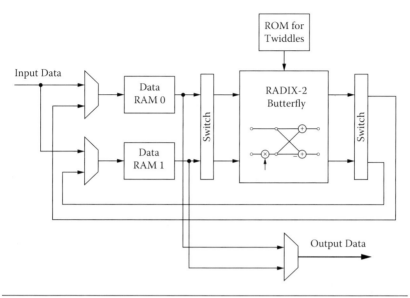

Figure 5.4   Block diagram of the Radix2-Burst I/O architecture. (From Hemphill, E. et al. 2007. Peak cancellation crest factor reduction reference design, Xilinx Application Note, December 5.)

architecture, separates the input data into two halves so the butterfly is smaller. This means it is smaller in size than the Radix4 solution. The fourth scheme is based on the Radix2 architecture. The Radix2-Lite-Burst I/O uses a time-multiplexed approach to the butterfly and the butterfly is even smaller but the transform time is longer. In this project, the Radix2-Burst I/O architecture as shown in Figure 5.4 is used, and the hardware resource requirement is less compared to the other algorithms used to prototype FFT (Yiqun et al., 2006).

The signal cannot be simultaneously loaded and unloaded like Radix4, Burst I/O architecture and the loading should be stopped during the calculation of the transform.

The point sizes can be from 8 to 65536 with a minimum of block memories, which are used in this algorithm. When the point size is equal or less than 1024, both the block memory and distributed memory can be used for data memories and phase memories.

5.6.2 Using AccelDSP Software to Prototype IFFT

AccelDSP is a synthesis tool that transforms a design in MATLAB into a hardware module. This module can be VHDL or Verilog code. This tool controls an integrated environment with other design tools such as MATLAB and Xilinx ISE tools. There is a browser in the GUI that shows the design hierarchy, the M-files, and the generated HDL source files. In this project, AccelDSP is used to generate the IFFT and FFT blocks. To guide the synthesis process, the design objects in the project explorer window are used. There are some parameters that should be defined here.

The type of Radix is one of the important parameters to consider when designing IFFTs. Here, the Radix2 algorithm is selected, which has been discussed before. The other parameter is the IFFT length that denotes the number of differential points in the IFFT. There is also an option for an I/O format. With the data I/O format option in AccelDSP GUI, input and output data can be initialized. Single buffering does not parallel any operations. Double buffering parallels the loading and unloading of frames of data. Natural Order I/O only applies to Single Fly architectures. Decimation algorithms will naturally have inputs or outputs in digit/bit reverse ordering. The decimation algorithm parameter will be set to decimation in time (DIT) or decimation in frequency (DIF) algorithm. DIF has natural order input and digit/bit reverse output; and DIT has digit/bit reverse input and natural order output. The Yes option will force input and output to be in natural order regardless of decimation type. The input data can be set to complex or real. The scaling is the 1/IFFT length ratio, which can be set. The Complex Multiplier is another option that chooses different complex multiplier architectures. Round Mode sets the Quantizer round mode property for all data path quantizers. If the Floor option is selected for round mode, the numbers between 0 and −1 will be rounded to 0 and all the other numbers, larger than 0 and less than −1, will round to the closest number. For example, −1.8 will be rounded to −2. There is also a section for input data width that shows the number of bits used to represent the input. Input Data Fract Width shows the number of bits used to represent the fractional part of the input word width. Twiddle Width is another parameter that shows the number of bits used to represent twiddle factors. Also, the twiddle factor width is the number of bits used to represent the fractional part of phase factors. The range of the phase factors is (−1, 1) and therefore 2 bits are always needed for the integer part of the phase factors. The data width can be also modified for the output of IFFT. Output data width depends on the scaling option. If the Scaling option is set to Yes, the output data width = input data width. If the Scaling is set to No, the output data width = input data width + log2 (IFFT length) + 1. The output data fractional width indicates the greater of either the input data fractional width or the twiddle fractional width.

5.6.3 Prototype of the Conventional Selected Mapping Method

Following the generation of the IFFT block with AccelDSP software and import as a MATLAB block in Simulink, the conventional selected mapping (C-SLM) method with $M = 2$ will be implemented. When $M = 2$, there are two multiplications in the block diagram of the C-SLM method. This multiplication block multiplies the input signal (X) and the phase sequence signal (B_i) together. Both inputs are complex, so two complex multipliers and two IFFT blocks are required. As mentioned earlier, the Xilinx blocksets should be used to implement the PAPR reduction method because only these blocksets can be compiled into the FPGA board. Since the complex multiplier is not available in the Xilinx blockset library, basic blocks shall be used to create complex multiplication. The complex multiplication can be rewritten as follows:

$$(B_iX_i = (B_i + jB_j)(X_i + jX_j) = (B_iX_i - B_jX_j) + j(B_iX_j + B_jX_i)$$

(5.10)

When the complex multiplication is rewritten as Equation (5.10), four basic multiplier blocks and one addition and one subtraction are required to represent the complex multiplier.

As Equation (5.10) shows, the imaginary parts and the real parts of the input signal and the phase sequence signal have to be separated prior to calculation. This is done by using a block named Complex to Real-Imag.

This block is available in the Simulink library but not in the Xilinx blocksets, which does not cause a problem because this separation can be performed before inserting the data into the C-SLM method, which should be compiled on the FPGA. In other words, the input signal that is in complex forms should be split into real and imaginary parts in order to be inserted to FPGA.

The System Generator block diagram for the complex multiplier is shown in Figure 5.5, which is designed based on Equation (5.10).

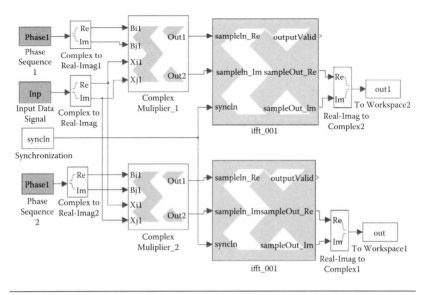

Figure 5.5 System Generator block diagram for the C-SLM method.

5.6.4 Implementation of the DSI–SLM Scheme

The System Generator block diagram for the DSI-SLM scheme is shown in Figure 5.5. The first step in the implementation of the DSI-SLM scheme is to generate dummy signals. Random signals based on the Gaussian series are used to generate dummy signals as shown by the block named DSI.

Two multiplexer blocks are designed to combine the dummy sequence with the input signal. They have a selector pin, which is connected to a relational block. The relational block is used to allocate the suitable location to the dummy sequence. The constant block is set to 201, which is the length of the input data before applying the dummy sequence. Then, the relational block compares the counter output with 201. When the counter output is less than 201, the selector pins of multiplexer blocks are set to 0 and when the counter output is higher than 201, the pins are set to 1. When the selector pins are set to 0, the multiplexer output will be the input data in pin *d0* and when the pins are set to 1, the outputs will select the input data of pin *d1*. As a result, the output data will be a combination of the input data and the dummy sequence.

To create DSI loops, the entire program must be repeated for the number of DSI loops denoted as *I*. Hence in implementation, each of the iteration runs and its PAPR is computed, which will be continued manually *I* times. It should be noted that the DSI-SLM program can be stopped at any time once the PAPR satisfies the condition and therefore the number of iterations might be less than *I*.

By finalizing the implementation of the DSI-SLM scheme, the "Netlist" is generated in order to create the VHDL code that can be compiled into the FPGA. In order to perform it, the System Generator block in the Xilinx library is used. The VHDL code of the DSI-SLM scheme is available in Appendix A.

Figure 5.6 presents the Co-Simulation block of hardware implementation by using JTAG. We have applied the input data and dummy generator and all the other components. We are now able to actually compile the program and test it through the FPGA board.

By running the JTAG Co-Simulation block, it will be compiled into the FPGA and the input samples will be imported to the FPGA and the output samples will be exported to the PC running with MATLAB. The maximum speed can be set to 4 MHz, which is sufficient for this demonstration. However, higher speeds can be achieved by using an Ethernet cable. The output samples from the FPGA will be captured and applied to extract the PAPR value.

It should be noted that the PAPR calculation is performed offline on the PC due to the complexity issue; however, in an actual scenario it is performed in digital signal processing.

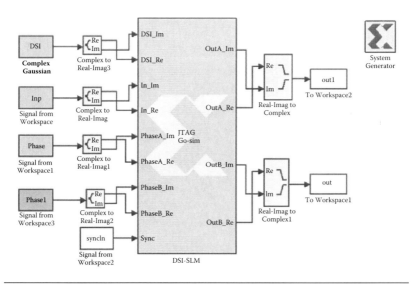

Figure 5.6 The hardware Co-Simulation of the DSI-SLM scheme.

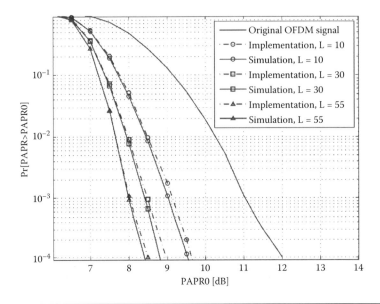

Figure 5.7 Comparison between simulation and implementation results for the DSI-SLM scheme, number of DSI iteration (I) = 10.

Figure 5.7 shows the PAPR performance comparison with about 0.1 dB difference between the MATLAB simulation and FPGA implementation result, which is acceptable due to the FPGA input resolution. This result is achieved when the number of iterations is 10 (I = 10). If the number of iterations is increased, meaning the same signal is combined with different types of dummy signals, hence the PAPR reduction will be enhanced because with different dummy signals, the probability of having less PAPR is increased.

As shown in Figure 5.7, the simulation and implementation produce almost similar results. When L = 55, the average PAPR of 8.5 dB is achieved, which has about a 3.5 dB reduction compared to the original OFDM signal at 10^{-4} CCDF.

The slight difference between simulation and implementation is due to the limitation of FPGA resolution, which is discussed in the next section.

5.6.5 Hardware Resource Consumption

Tables 5.6, 5.7, and 5.8 present the hardware resource consumption of the DSI-SLM, DSI, and C-SLM (M = 2) methods, respectively. This analysis is called the device utilization summary. To ensure

Table 5.6 Hardware Resource Consumption of DSI-SLM Scheme

XC5VFX30T-1FF665	RESOURCES USED/AVAILABLE	PERCENTAGE OF CONSUMPTION
Slices	4,341/20480	21%
DSP48 slices	26/64	40%
Number of fully used LUT-FF pairs	2340/4179	56%
IO Utilization units	170/360	47%

Table 5.7 DSI Method Hardware Resources

XC5VFX30T-1FF665	RESOURCES USED/AVAILABLE	PERCENTAGE OF CONSUMPTION
Slices	2,358/20480	11%
DSP48 slices	9/64	14%
Number of fully used LUT-FF pairs	2,241/4179	56%
IO Utilization units	86/360	23%

Table 5.8 C-SLM Hardware Resources When $M = 2$

XC5VFX30T-1FF665	RESOURCES USED/AVAILABLE	PERCENTAGE OF CONSUMPTION
Slices	3996/20480	19%
DSP48 slices	30/64	46%
Number of fully used LUT-FF pairs	3825/4179	91%
IO Utilization units	142/360	39%

that the design will fit in the selected device, looking at the design utilization summary section of the design summary report in ISE environment is essential. It reports how many percentages of available hardware resources are required for implementing the method. To obtain this table, the "NGC Netlist" file has to be generated by the System Generator. This file includes the method structure, and the ISE software is able to analyze it and extract the hardware resource for the design.

As shown in Tables 5.9 and 5.10, when M increases from 2 to 4, the hardware resources also increase significantly due to the number of IFFTs. Observe that for $M = 16$, these values will be increased more than 16 times.

Table 5.10 shows a comparison between the DSI-SLM, C-SLM, and DSI methods in terms of hardware resource consumption.

Table 5.9   C-SLM Hardware Resources When $M = 4$

XC5VFX30T-1FF665	RESOURCES USED/AVAILABLE	PERCENTAGE OF CONSUMPTION
Slices	8,115/20480	39%
DSP48 slices	60/64	93%
Number of fully used LUT-FF pairs	7,796/4179	186%
IO Utilization units	287/360	79%

Table 5.10   Hardware Resource Comparison

XC5VFX30T-1FF665	DSI-SLM CONSUMPTION ($M = 2$)	C-SLM CONSUMPTION ($M = 4$)	DSI CONSUMPTION
Slices	22%	39%	11%
DSP48 slices	46%	93%	14%
Number of fully used LUT-FF pairs	25%	55%	56%
IO Utilization units	53%	79%	23%

Table 5.11   Estimated Power Summary of DSI-SLM (Ambient Temperature = 33°C)

		RESOURCES		
NAME	POWER (W)	USED	TOTAL AVAILABLE	UTILIZATION (%)
Logic	0.00514	5292	20480	25.8
Signals	0.01000	9296	—	—
DSP	0.00146	26	64	40.6
Total quiescent power	0.62417			
Total dynamic power	0.05610		N/A	
Total power	0.68027			

The DSP48 and I/O Utilization units of the C-SLM are used 93% and 79%, respectively, so implementation of the C-SLM method with $M = 8$ and $M = 16$ is not feasible due to the lack of hardware resources for the FPGA device.

The power consumed by the implemented DSI-SLM scheme is estimated by the ISE XPower analyzer, Xilinx tool, after the place and route process. The processor consumes a total power of 0.68027 watt and dynamic power of 0.05610 watt. Table 5.11 presents the details of the power report.

Based on the timing report, the data path delay of the C-SLM ($M = 4$) and DSI-SLM ($M = 2$) methods are 11.716 ns and 12.551 ns, respectively. The C-SLM method consumes a total power of 0.67572 watt and dynamic power of 0.05195 watt. Table 5.12 shows the details of the power consumption report.

The ISE tool is also able to generate an intellectual property (IP) block of the implemented program. The IP of the DSI-SLM scheme is shown in Figure 5.8. As shown in Figure 5.8, there are 11 input pins and 4 output pins. The inputs are mainly for imaginary and real parts of the input OFDM signal, the random phase sequences, and the dummy sequence. The output pins are real and imaginary parts of the candidate signals. The characteristics of the inputs and outputs are

Table 5.12   Estimated Power Summary of C-SLM, $M = 4$ (Ambient Temperature $= 33°C$)

		RESOURCES		
NAME	POWER (W)	USED	TOTAL AVAILABLE	UTILIZATION (%)
Logic	0.00416	4877	20480	23.8
Signals	0.01026	8091	—	—
DSP	0.00146	26	64	40.6
Total quiescent power	0.62377			
Total dynamic power	0.05195		N/A	
Total power	0.67572			

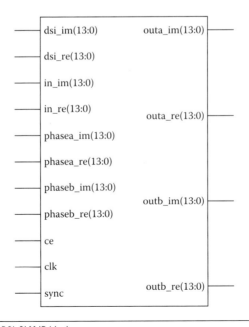

Figure 5.8   The DSI-SLM IP block.

Table 5.13 DSI-SLM Pin Characteristics

PIN NAME	DESCRIPTION
dsi_im (13:0)	Imaginary part of the dummy signal, Signed 14/12 bits
dsi_re (13:0)	Real part of the dummy signal, Signed 14/12 bits
in_im (13:0)	Imaginary part of the input signal, Signed 14/12 bits
in_re (13:0)	Real part of the input signal, Signed 14/12 bits
phasea_im (13:0)	Imaginary part of the phase A, Signed 14/12 bits
phasea_re (13:0)	Real part of the phase A, Signed 14/12 bits
phaseb_im (13:0)	Imaginary part of the phase B, Signed 14/12 bits
phaseb_re (13:0)	Real part of the phase B, Signed 14/12 bits
sync	Synchronization pin, Unsigned 1/0
outa_im (13:0)	Imaginary part of the output signal A, Signed 14/12 bits
outa_re (13:0)	Real part of the output signal A, Signed 14/12 bits
outb_im (13:0)	Imaginary part of the output signal B, Signed 14/12 bits
outb_re (13:0)	Real part of the output signal B, Signed 14/12 bits
ce	Clock enable pin
clk	Clock pin of the DSI-SLM block

also presented in Table 5.13. According to this table, dsi, in, phasea, phaseb, outa, and outb ports include imaginary and real parts with a 14 bit length and 12 bit floating point, respectively.

The resolution of 14/12 can be increased, which requires a matching design in the IFFT block. This means that the IFFT should also be designed according to the resolution that is used in the whole system.

Higher resolution consumes a higher number of I/O utility units in FPGA, which means that the hardware consumption will be increased. Here, to avoid a hardware resource consumption mismatch, the resolution is set to 14/12, which is the same as the IFFT design. This resolution means that the fixed point signal has 12 bits fractional parts and 1 bit is for the integer and 1 bit is for the sign, which together becomes 14 bits length. The two other pins are clk and ce, which are used to determine the clock of the DSI-SLM block. The other pin characteristics of the DSI-SLM block are explained in Table 5.13.

5.7 FPGA Implementation of the Optimum Phase Sequence with the Dummy Sequence Insertion Scheme

Another method is also proposed to increase the performance and flexibility of the DSI-SLM scheme. This proposed method is called the optimum phase sequence with dummy sequence insertion

(OPS-DSI) scheme. This method has less complexity with about 0.5 dB better PAPR performance. In this section, the hardware implementation of the OPS-DSI scheme is presented, which uses the same platform and FPGA setup as the DSI-SLM scheme.

5.7.1 Implementation of the OPS-DSI Transmitter

A block diagram of the OPS-DSI scheme was presented in Figure 4.40. Because the OPS-DSI scheme outperforms DSI-SLM, the complete implementation of this scheme is also presented here for both a transmitter and receiver.

It should be noted that the receiver implementation of DSI-SLM is almost the same as OPS-DSI. The System Generator block diagram of the OPS-DSI scheme is shown in Figure 5.9. The m.file of initialization process is provided in Appendix B. As seen in Figure 5.9, there is only one IFFT block for the OPS-DSI scheme and so the complexity is significantly reduced. According to the OPS-DSI block diagram, following the serial-to-parallel converter block there is a multiplication operator. The implementation of this complex multiplier is similar to the DSI-SLM and C-SLM methods as explained in the previous sections. The input signal should be multiplied with a random phase sequence.

To explicitly explain the procedure of the implementation of OPS-DSI, we should note that the input signal to the FPGA should be in a sample-based format, meaning that the input signal has to be one dimensional before inserting it into FPGA. Since the phase sequence matrix (B) is not one dimensional, it has to be reshaped to a one-dimensional vector. In order to make the iterations of the OPS, which are denoted with P, the input data symbols are multiplied with the M phase sequence and then this process is repeated for several numbers of samples (Z).

The result is reshaped into a $1 \times [(256 - L) \times Z \times M \times I]$ matrix. This data stream is introduced to the dummy insertion block. In the dummy insertion block, the input data is multiplexed with the dummy sequence, which will also be converted to a one-dimensional vector as $1 \times (L \times Z \times I \times M)$ and the IFFT will be performed.

Similar to the DSI-SLM scheme, there are gateway blocks to transfer the signal outside of the FPGA. The real and imaginary parts are converted to the complex form and the PAPR is computed. When the PAPR is computed, it is given to a threshold comparator block.

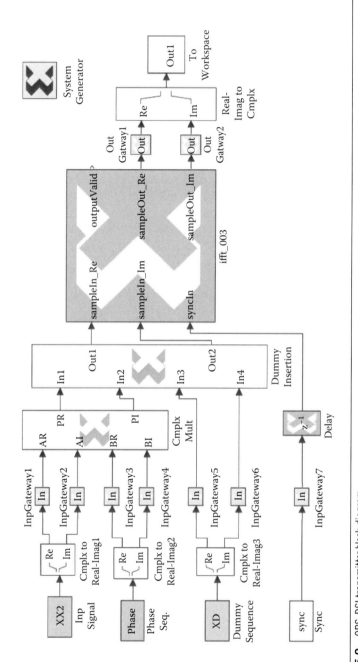

Figure 5.9 OPS–DSI transmitter block diagram.

If the PAPR is less than the predefined threshold value, which is 7 dB here, the signal will be transmitted. The threshold value is chosen because of the standard definition. According to the Worldwide Interoperability for Microwave Access (WiMAX) standard (IEEE STD 802.16–2005), the PAPR value should be more than 7 dB.

When the design is completed, the "Co-Sim" block can be generated using the System Generator. This block can be directly compiled to the FPGA using the System Generator tools and then the input data can be introduced to the implemented algorithm and the output signal can be captured from the "OutGateway" pins as shown in Figure 5.10.

The result presented in Figure 5.11 is based on when the DSI iteration number is set to 10 ($I = 10$). It should be noted that the inner loop iteration will be reset when the second loop is executed. If both iterations are performed and the PAPR of the signal is still higher than the threshold, the signal with the minimum PAPR will be transmitted regardless of the PAPR threshold value.

In Figure 5.11, the CCDF result of the simulation of the OPS-DSI technique is compared with the implementation result of the OPS-DSI. The simulation is performed when $P = 4$ and $I = 10$. According to Figure 5.11, the CCDF result of implementation is slightly less than the simulation, which is expected for the FPGA because of the resolution of the data. By increasing the I/O bit resolution, the PAPR performance of implementation overlaps the simulation results.

The result in Figure 5.12 is for the OPS-DSI technique when $P = 2$ and $S = 10$. It means that the phase sequence multiplication loop is repeated twice. When the OPS loop is repeated, everything within its loop is also repeated including the loop for the dummy insertion.

In the simulation of the OPS-DSI scheme, at any time instant that the PAPR passes the threshold comparator, the program will be stopped and the signal is sent to the amplification stage.

However, in the implementation, the procedure is different in that the signal must be in a one-dimensional format and the multiplication and IFFT are performed based on a sample-based signal. The implementation of OPS-DSI is carried out without any loop and the samples are consequently constructed.

According to Figure 5.12, the implementation result of the PAPR reduction performance is slightly less than the simulation result.

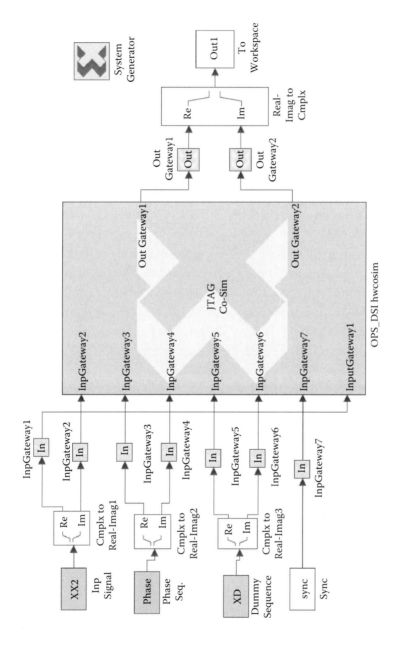

Figure 5.10 System Generator block diagram of the OPS-DSI Co-Sim block.

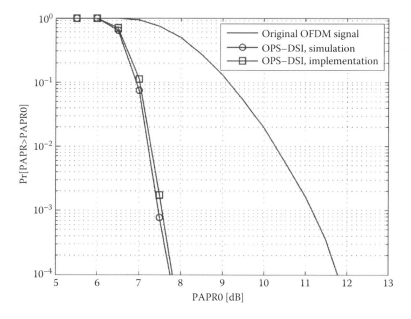

Figure 5.11   The PAPR performance comparison between implemented OPS-DSI and simulated OPS-DSI when $P = 4$.

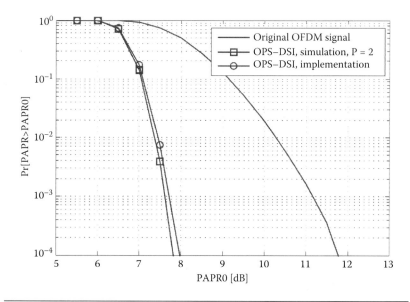

Figure 5.12   The PAPR performance comparison between implemented OPS-DSI and simulated OPS-DSI when $P = 2$.

This is completely acceptable because the result achieves the minimum PAPR reduction according to the IEEE standard, which is on average 3.5 dB.

5.7.2 *Implementation of the OPS-DSI Receiver*

To verify the BER performance of the OPS-DSI scheme, a design of the receiver part is required. Figure 5.13 shows the System Generator block diagram of the OPS-DSI receiver. The signal after transmission will go through the channel and will be received at the receiver, which after the low noise amplifier (LNA), downconversion, and analog-to-digital (A/D) converter, will be converted to the real and imaginary parts in the FPGA. The real and imaginary signal in the time domain will then convert to the frequency domain by using the FFT block. The real part of the FFT output is saved in a RAM that is shown by RAM_REAL in Figure 5.13. The imaginary part of the FFT output is saved in a RAM that is shown by RAM_IMAG.

As shown in Figure 5.13, a single port RAM is used in this program. Single port RAM is one of the components from the Xilinx blockset library in Simulink. It has one output port and three input ports for address, input data, and write enable (WE). When WE = 0, the output port has the value at the location specified by the address port. When WE = 1, the mode "no read on write" is selected.

The counter block is connected to the RAM_REAL block and counts from 1 to 201. So the data located between 1 and 201 are transferred to the output and the dummy sequence is discarded. The RAM_IMAG works in the same way and the output of these two memories are connected to the complex division block. It should be noted that the side information is extracted from the received signal offline, which is located at samples 252 to 256. The optimum phase sequence will be generated and inserted to the input port that is indicated by phase in Figure 5.13.

The real and imaginary parts of the optimum phase sequence are connected to the PHASE_SEQ_RAM. This RAM has a dual port memory. Both outputs of the PHASE_SEQ_RAM are connected to the Complex Division block.

In the Complex Division block, the input signal is divided by the phase sequence. The details of complex division are explained later.

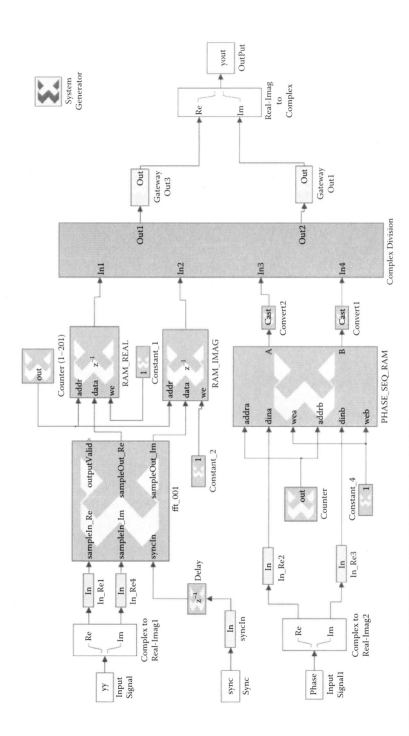

Figure 5.13 OPS-DSI receiver implementation.

The real and imaginary parts of the original signal are passed through the gateway block and then converted to the complex format. In this stage, the original data is extracted. Now the BER results should be generated to verify the operation of the OPS-DSI implementation in FPGA.

As the simulation is not performed in real time, hence the BER measurement is also performed offline. The data is first transmitted and stored in the lookup table (LUT) and by passing it through the additive white Gaussian noise (AWGN) channel, the result will be applied as the input for the receiver. By running the implementation of the receiver in FPGA, the output data is captured. The BER can be demonstrated as the comparison of the transmitted symbols and the received signals.

The BER of the OFDM signal with the OPS-DSI scheme is measured in two conditions: when $P = 2$ and $P = 4$. The BER performance is compared with the BER measured in simulation. The comparison is shown in Figure 5.14. The BER performance result of OPS-DSI implementation is slightly inferior compared with the simulation result and the original OFDM signal.

As mentioned earlier, the main contribution in the receiver of the OPS-DSI is in the complex division. The received signal has a

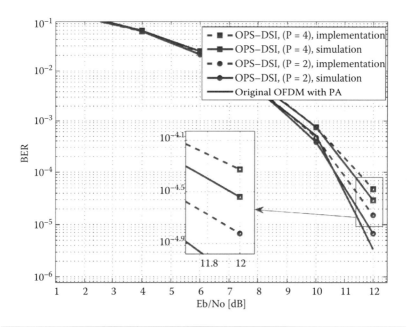

Figure 5.14 BER performance comparison between implementation and simulation.

complex format, which should be divided by a phase sequence, which is also in a complex format. The design of this complex division is explained in the next section.

The IP of the OPS-DSI scheme is shown in Figure 5.15, which is generated by ISE tools. The information regarding pin characteristics are presented in Table 5.13. According to Figure 5.15, the pins of the IP can be recognized. As explained in Table 5.13, the pins of Inpgateway1, Inpgateway2, Inpgateway3, and Inpgateway4 are real and imaginary parts of input signal and phase sequence, respectively.

It is shown in Figure 5.15 that the pins of Inpgateway5 and Inpgateway6 are used for real and imaginary parts of the dummy sequence. As presented in Table 5.14, the pin of Inpgateway7 is used for synchronization, which determines the length of the signal. The length of the signal is set to 14 bits with 12 floating points.

The synchronization processes as well as the clocking process shown by the clk and ce pins are the same as DSI-SLM IP.

The power consumed by the OPS-DSI scheme is estimated by the ISE XPower analyzer, Xilinx tool, after the place and route process. The processor consumes a total power of 0.66526 watt and dynamic power of 0.04241 watt. Table 5.15 shows the details of the power consumption report.

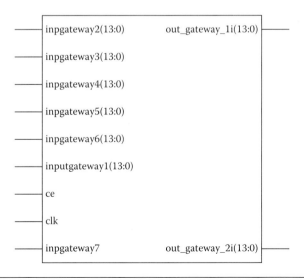

Figure 5.15 OPS-DSI block.

Table 5.14 OPS-DSI Block Characteristics

PIN NAME	DESCRIPTION
Inpgateway1 (13:0)	Real part of the input signal, Signed 14/12 bits
Inpgateway2 (13:0)	Imaginary part of the input signal, Signed 14/12 bits
Inpgateway3 (13:0)	Real part of the phase sequence, Signed 14/12 bits
Inpgateway4 (13:0)	Imaginary part of the phase sequence, Signed 14/12 bits
Inpgateway5 (13:0)	Real part of the dummy sequence, Signed 14/12 bits
Inpgateway6 (13:0)	Imaginary part of the dummy sequence, Signed 14/12 bits
Inpgateway7 (13:0)	Synchronization pin
Out_gateway_1i(13:0)	Real part of the output signal, Signed 14/12 bits
Out_gateway_2i(13:0)	Imaginary part of the output signal, Signed 14/12 bits
ce	Clock enable pin
clk	Clock pin of the DSI-SLM block

Table 5.15 Estimated Power Summary of OPS-DSI (Ambient Temperature = 33°C)

		RESOURCES		
NAME	POWER (W)	USED	TOTAL AVAILABLE	UTILIZATION (%)
Logic	0.00234	2745	20480	13.4
Signals	0.00582	4988	—	—
DSP	0.00051	13	64	20.3
Total quiescent power	0.62285			
Total dynamic power	0.04241		N/A	
Total power	0.66526			

According to the timing report, the total data path delay of OPS-DSI is 10.937 ns, which is less than C-SLM and DSI-SLM. It should be noted that less path delay results in a higher data rate and spectrum efficiency. The main reason that a high delay causes data rate degradation is that a long delay decreases the bandwidth (Bosch and Gatti, 1989; Katz, 2004), which according to the Shannon-Hartley theorem (Hartley, 1928) has a direct relation with the data rate. The condition is as follows:

$$BW \leq \frac{1}{4\Delta t_s} \tag{5.11}$$

where Δt_s is the total path delay and BW is the signal bandwidth.

As an example, if the total path delay is 10 ns, the maximum bandwidth can be 25 MHz. It should be noted that Δt_s includes the power amplifier delay, which normally should be less than 5 ns.

5.8 Implementation of Complex Division in the Receiver

If there is no divider block inside the FPGA then this block should be designed by the other existing blocks like the multiplier, adder, and subtracter. The implementation is expected to be a part of a baseband processor with a small size and should be able to handle the high-speed requirements.

There are two ways to perform a division with logical circuitry. The way that is used by the Xilinx CoreLib is to calculate each bit exactly. It is based on the CORDIC algorithm. The other method is based on the Newton-Raphson method. The following section describes the complex divider algorithm and then the subsections for implementing it are explained. Most of the study for implementing the division is based on real-valued operands. Very little research has focused on complex division (Agrawal and Khandelwal, 2006; Alfredsson, 2005; Ercegovac and Muller, 2003; Obermann and Flynn, 1997). The complex divider that is designed here is actually a combination of previous works (Agrawal and Khandelwal, 2006; Alfredsson, 2005; Ercegovac and Muller, 2003), but none of them develop the complete divider as a block. In Agrawal and Khandelwal (2006), an initial approximation technique was designed and in Ercegovac and Muller (2003) a technique for prescaling the input values was introduced. Alfredsson (2005) proposed a Newton method. Here, various techniques were tested and examined, and then the CORDIC algorithm also applied for the divider and compared to the complex divider. The complete divider was found to have many advantages over the other methods. Complex division in the Cartesian format can be performed by a combination of real-valued additions, multiplications, and divisions. The division between the complex numbers $x = a + bi$ and $y = c + di$ can be done as follows (Alfredsson, 2005):

$$\frac{y}{x} = \frac{c+di}{a+bi} = \frac{(c+di)(a-bi)}{(a+bi)(a-bi)} = \frac{(c+di)(a-bi)}{(a^2+b^2)} = \frac{(ca+db)+(ca-da)i}{(a^2+b^2)}$$

$$(5.12)$$

A modified type of the formula can be found in Ercegovac and Muller (2003). That method, however, requires an extra division and is not used in this book. If there are several numbers, xi that should be divided with one single number y; it is better to calculate the reciprocal of x, $z = 1/x$, and then multiply z with xi to get the result. Only one inversion is then required. The calculation of $z = 1/x$ is carried out as follows:

$$\frac{1}{a+bi} = \frac{a-bi}{(a+bi)(a-bi)} = \frac{a-bi}{a^2+b^2} = \frac{a}{a^2+b^2} - \frac{b}{a^2+b^2} \cdot i \quad (5.13)$$

This requires two multiplications, one addition, and two divisions.

In the Newton-Raphson method for division, the accuracy of the initial approximation is crucial as it can cause faster convergence to the final result.

The division in general can be shown as

$$Q = \frac{N}{D} \quad (5.14)$$

where Q is the quotient, N is the numerator (dividend), and D is the denominator (divisor). In this formula, N and D are assumed to be of the form

$$N = 1.x_1 x_2 x_3 \ldots\ldots x_k$$
$$D = 1.y_1 y_2 y_3 \ldots\ldots y_k \quad (5.15)$$

Here, the Newton-Raphson method is applied for the divider, which is simpler and has less hardware resources.

5.8.1 Newton-Raphson Division

Newton-Raphson division is a well-known iterative method that finds the root of a nonlinear function. Assume the function $g(x)$ and let u be a root of the equation $g(x) = 0$; first start with x_0, which is a good estimate of u and $u = x_0 + l$. The number l measures how far the estimate x_0 is from the truth. Since l is small, the linear approximation can be used to conclude that (Ercegovac and Muller 2003)

$$0 = g(u) = g(x_0 + l) \approx g(x_0) + l g'(x_0) \quad (5.16)$$

Then it can be concluded that

$$u = g_0 + l \approx g_0 - \frac{g(x_0)}{g'(x_0)} \quad (5.17)$$

If x_i is the current estimate, then the next estimate will be x_{i+1} given by

$$x_{i+1} = x_i - \frac{g(x_i)}{g'(x_i)} \tag{5.18}$$

Equation (5.18) is called the Newton-Raphson formula. In order to compute the reciprocal, the following function and its derivative are used. It can also be written as

$$x_{i+1} = x_i(2 - Dx_i) \tag{5.19}$$

Using the form in Equation (5.19), one square, one multiplication, one shift, and one subtraction are required for the computation of x_{i+1}. In the implementation of the complex divider, the Newton-Raphson method is applied because of its accuracy and simplicity.

5.8.2 Error Analysis

To analyze the divider results and see the accuracy of this technique, the error should be considered.

Let $\varepsilon_i = \dfrac{1}{D} - x_i$ be the error at ith iteration, then (Ercegovac and Muller, 2003)

$$\varepsilon_{i+1} = \frac{1}{X} - x_{i+1} = \frac{1}{D} - x_i(2 - x_i D) \tag{5.20}$$

This can also be expressed as (Ercegovac and Muller, 2003):

$$\varepsilon_{i+1} = D(1/D - x_i)^2 = D\varepsilon_i^2 \tag{5.21}$$

Equation (5.21) clearly shows that the absolute error decays quadratically in each iteration.

5.8.3 Initial Approximation Techniques

An important part of the Newton-Raphson method is an initial approximation in the iteration formula. The number of necessary iterations depends on the accuracy of the initial approximation. With the reduction in the number of iterations, not only will the area of the design decrease but it also helps to reduce the delay. Thus, accurate initial approximation is very important. Various techniques are available to calculate the initial approximation, some of which include a linear approximation, direct lookup table, and lookup table followed by multiplication (Agrawal and Khandelwal, 2006).

5.8.4 Hardware Structure of the Complex Divider

There are mainly four steps that should be performed. The first is sign adjustment and prescaling of the inputs into the required range of the dividends. The second step is to calculate the divisor and change the scale to the proper range. The third step is to perform the actual division. The fourth step is postscaling the result.

A graphical representation of the structure can be viewed in Figure 5.16. However, the divider that is designed here is capable of doing the four steps, but because the dividend or numerator is 1, the first step is skipped and only three steps remain for implementing the divider.

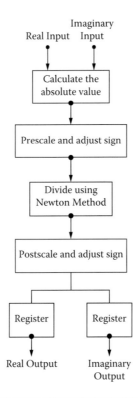

Figure 5.16 The flowchart for the complex divider implementation.

In Figure 5.16, one block handles the calculations on the real part and the other one on the imaginary part. As the values are absolute after the first step, there is no need to take care of the sign of the values. This creates fewer hardware resources. Next, each part is explained separately.

5.8.5 Divisor Scaling

The input of the block is a set of 14 bits, two fixed point numbers in the range (–1, 1). The first thing that should be done is to remove the sign, but in the absolute of the input signals there is no negative sign. The range is then between (0, 2). The divider requires both inputs to be in the range (1, 2). As the inputs are not in the range which is required, then it has to be prescaled into the proper range and the output postscaled to return to the previous values. The prescaling is carried out by finding the first nonzero bit, starting from the leftmost position. The value is then shifted to the left until the nonzero bit is found. Then the range will be (0.5, 1). But the values should be in the range (1, 2). Now the fractional point is moved one step to the right. The shift will be compensated in the postscaling step. The Xilinx block for implementing this part is shown in Figure 5.17.

As shown in Figure 5.17, there are three inputs: din is the divisor that should be prescaled, cin is the number of the left shift that should be stored in the register for the next stage, and msb is the most significant bit and it will keep the left msb until it finds one (Ercegovac and Muller, 2003). According to Figure 5.17, $x(i)$ is the initial guess and based on trial and error is assumed to be 0.7, and because the input values are prescaled and are between 1 and 2, there is nothing else to be done. Otherwise finding the initial value depends on the input data, but here it has been found that this value is a good approximation.

5.8.6 Newton–Raphson Method

The Xilinx simulation of the Newton–Raphson method is shown in Figure 5.18. As discussed earlier, the Newton–Raphson method is performed for one iteration.

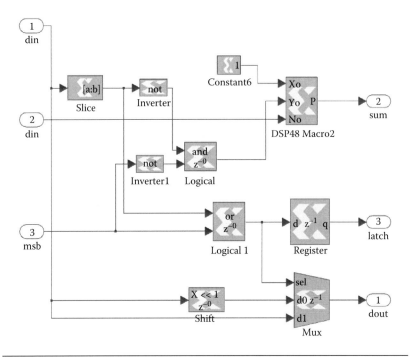

Figure 5.17 The Xilinx blocksets of the standard input/dividend prescaling block.

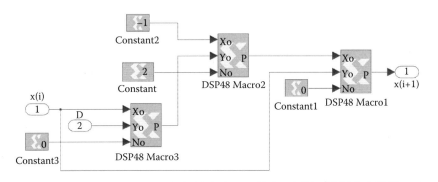

Figure 5.18 The Newton-Raphson implementation with Xilinx blocksets.

5.8.7 *Postscaling of Division Values*

As mentioned earlier, the output needs to be scaled. If the dividend N is scaled by $2r$ and the divisor D by $2l$, the division result becomes as follows (Ercegovac and Muller, 2003):

$$u = \frac{N.2^r}{D.2^l} = \frac{N}{D} \cdot \frac{2^r}{2^l} = \frac{N}{D} .2^{r-l} \qquad (5.22)$$

Then, the output U can be calculated as

$$U = \frac{N}{D} = u.2^{l-r} \tag{5.23}$$

That is, u should be scaled by 2^{r-l}.

Figure 5.19 shows the complex divider block, which is created after the completion of the design. As can be seen in this block, there are two inputs that are the complex I and Q signals and the outputs ID and QD, which are the division signals of the inputs.

Tables 5.16 and 5.17 present the hardware resource consumption of the CORDIC algorithm of the complex division and the design complex divider, respectively. It can be observed that the new algorithm performs better compared to the CORDIC algorithm.

Complex Divider Block

Figure 5.19   The complex divider block.

Table 5.16   Hardware Resources for the CORDIC Divider

XC5VFX30T-1FF665	RESOURCES USED	CONSUMPTION (%)
Slices	875	17
DSP48 Slices	792	3.8
Number of fully used LUT-FF pairs	484	92

Table 5.17   Hardware Resources for the Divider Implementation

XC5VFX30T-1FF665	RESOURCES USED	CONSUMPTION (%)
Slices	258	5
Flip Flops	273	1.33
DSP48 Slices	484	32.8
Number of fully used LUT-FF pairs	404	75

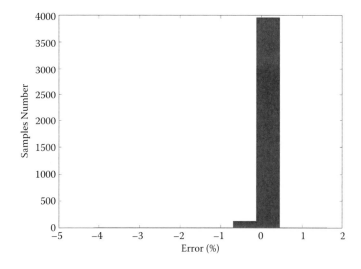

Figure 5.20 The error graph of the CORDIC divider for 4096 samples.

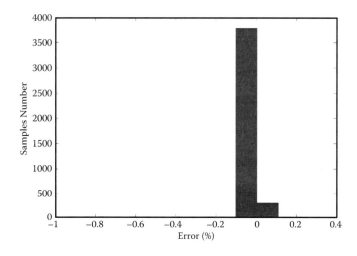

Figure 5.21 The error graph of the complex divider for five iterations of Newton-Raphson and 4096 samples.

Figures 5.20 and 5.21 show the histogram plot of the error analysis of the divider for 4096 samples. As shown in Figure 5.20, the error percentage for the CORDIC divider is more than the error percentage in Figure 5.21, which is for the complex divider that has been designed. These graphs also show the advantages of the complex divider that has been designed here. In Figure 5.20, the error percentage is more within −1% and 1%. These error values are between the actual divider result

Table 5.18   Hardware Resource Consumption Comparison of OPS-DSI and DSI-SLM Schemes

XC5VFX30T-1FF665	OPS-DSI USED HARDWARE	OPS-DSI CONSUMPTION	DSI-SLM CONSUMPTION
Slices	2360	11%	21%
DSP48 Slices	13	20%	40%
Number of fully used LUT-FF pairs	2304	57%	56%
IO utilization	114	31%	47%

and the result of implementation with Xilinx blocks. The error percentage should be very low to have an accurate division.

According to Figure 5.21, the amount of the error percentage is more between −0.1% and 0.1%, which is much smaller than the CORDIC algorithm. The difficulty of this complex division is when the input signal values are very small, in the range of 0.00001, and the inverse of these values is so difficult to find, and a large number of iterations in the divider are required. This is the main reason that this error percentage is high for some samples. It can be seen that the error percentage is scattered in other parts of the histogram but just for those samples that have very small values.

5.9 Hardware Resource Consumption of the OPS-DSI Scheme

The hardware resource consumption of the OPS-DSI is presented in Table 5.18. This table is generated by using Xilinx ISE tools. As shown by the table, the proposed OPS-DSI scheme consumes fewer hardware resources. Since the number of multiplications and additions required for implementation of an algorithm is directly related to the number of DSP48 slices, the main consideration in hardware resource consumption is the percentages of used DSP48 slices.

As shown in Table 5.18, the OPS-DSI scheme has consumed only 20% of the DSP48 slices compared to the DSI-SLM scheme. It can be observed that the hardware resource consumption of OPS-DSI outperforms DSI-SLM.

5.10 Summary

In this chapter, the hardware implementation of the DSI-SLM and OPS-DSI schemes were presented. The Xilinx FPGA kit was used to carry out the implementation. MATLAB, ISE, and AccelDSP software were used to create the test platform. An input signal based on IEEE 802.16e was designed and given to the DSI-SLM and OPS-DSI schemes, then the dummy insertion block and the C-SLM block were designed using the Xilinx Blockset library in MATLAB. In the dummy insertion block, multiplexer blocks are used and complex multipliers have been designed for phase sequence multiplication. The comparison of simulation and implementation results presented a slight difference that can be compensated by increasing the FPGA input output bit resolution.

The timing and power consumption of the OPS-DSI scheme outperform C-SLM and DSI-SLM.

The receiver of the OPS-DSI is designed and implemented in the FPGA. A novel complex divider is used to perform the division that is required to extract the original data stream from the received signal. Then the BER is measured and it is compared with the simulation results.

References

Agrawal, G., and Khandelwal, A. 2006. A Newton-Raphson divider based on improved reciprocal approximation algorithm. University of Texas at Austin.

Alfredsson, E. 2005. Design and implementation of a hardware unit for complex division. Master of science dissertation, Linköping University, Sweden.

Baxley, R. J., and Zhou, G. T. 2007. Comparing selected mapping and partial transmit sequence for PAR reduction. *IEEE Transactions on Broadcasting* 53(4):797–803.

Bosch, W., and Gatti, G. 1989. Measurement and simulation of memory effects in predistortion linearizers. *IEEE Transactions on Microwave Theory and Techniques* 37(12):1885–1890.

Ercegovac, M. D., and Muller, J.-M. 2003. Complex division with prescaling of operand. Proceedings of IEEE International Conference on Application-Specific Systems, Architectures, and Processors (ASAP 2003), Hague, Netherlands, June 24–26.

Hartley R. V. L. 1928. Transmission of Information. *Bell System Technical Journal.*

Hemphill, E., Summerfield, S., Wang, G., and Hawke, D. 2007. Peak cancellation crest factor reduction reference design, Xilinx Application Note, December 5.

Institute of Electrical and Electronics Engineers. 2005. IEEE STD 802.16e–2005, IEEE Standard for Local and Metropolitan Area Networks.

Institute of Electrical and Electronics Engineers. 2011. IEEE 802.16e WiMAX OFDMA Signal Measurements and Troubleshooting; Agilent Application Note 1578.

Katz, A. 2004. Linearizing high power amplifiers. http://www.lintech.com/PDF/hpa.pdf.

Obermann, S. F., and Flynn, M. J. 1997. Division algorithms and implementations. *IEEE Transactions on Computers* 46(8):833–854.

Ryu, H. G., Lee, J. E., and Park, J. S. 2004. Dummy sequence insertion (DSI) for PAPR reduction in the OFDM communication system. *IEEE Transactions on Consumer Electronics* 50(1):89–94.

Shanblatt, M. A., and Foulds, B. 2005. A Simulink-to-FPGA implementation tool for enhanced design flow. Proceedings of IEEE International Conference on Microelectronic Systems Education, 2005 (MSE '05), California, June 12–14.

Yiqun, Z., Hayes-Gill, B. R., Morgan, S. P., and Hoang, N.C. 2006. An FPGA based generic prototyping platform employed in a CMOS laser Doppler blood flow camera. IEEE International Conference on Field Programmable Technology (FPT 2006), Bangkok, Thailand, December 13–15.

6

POWER AMPLIFIER
LINEARIZATION

6.1 Introduction

Power amplifiers (PAs) are necessary components in communication systems and they are nonlinear in a certain operating region. Power amplifier nonlinearity results in spectral broadening, which leads to adjacent channel interference (ACI) and superfluous radiation. One way to avoid these impacts is to back off the power amplifier from its saturation point or from the nonlinear region, but this causes degradation in power amplifier efficiency that is not desired (typically less than 10%) (Wright, 2002). More than 90% of the direct current (DC) power is lost and turns into heat, hence other methods need to be investigated in which the spectrum broadening can be suppressed while the power efficiency is maintained. In the new broadband communication technologies, the power amplifier exhibits more severe nonlinearity. The transmission formats, such as wideband code division multiple access (WCDMA) and orthogonal frequency division multiplexing (OFDM), have high peak-to-average power ratios (PAPRs), which means large fluctuations in their signal envelopes. By increasing the number of base stations and then the number of power amplifiers, an improvement in efficiency of the power amplifier reduces the cost of the system. To enhance the power amplifier efficiency and maintain linearity of the power amplifier, a linearization technique is pivotal to combat nonlinearity.

It is important to note that by increasing the number of users, a greater amount of bandwidth is required. One of the effective methods used to increase bandwidth is by applying diversity techniques, which have been applied in the current 3G (third generation) standard specifications (Vuolvei, 2003). This technique is called a multiple input multiple output (MIMO) system. But, with each additional antenna, an extra transceiver has to be embedded, which can significantly increase the system cost. Another method is the digital predistortion (DP) technique. The DP technique overcomes the nonlinearity of the PAs by increasing the power efficiency, which reduces the cost of the overall system (Kenington, 2000). DP among all linearization techniques is the one that is low cost and with high efficiency and also high flexibility. By applying DP, which is implemented in the baseband of the communication system, the nonlinearity of the power amplifier is reduced and it allows the use of a high power amplifier with high efficiency in the systems.

Another important fact in studying PAs is the memory effects, which are the main issue of this book. The focus here is on the short-term memory effects, which cause the characteristics of the PA to vary with time. This effect is more important when the high bandwidth signals are applied. The memory effects cause an increase in adjacent channel leakage ratio (ACLR) and error vector magnitude (EVM). The other factor that might cause problems to the performance of the predistortion is the effect of noise. Noise here can be the result of an analog part such as a digital-to-analog converter (DAC), mixer, and so on, which in this book are not considered, because the predistortion here is implemented in the baseband. The other category of memory effects is electrical or short-term effects that cannot be compensated by a simple feedback. In this book, the linearization techniques of the power amplifiers of mainly class AB are investigated, then the DP technique is chosen as it is the most cost effective and most efficient among all the linearization techniques. The class AB power amplifier is chosen because it has more linearity as compared to other classes. A new technique has been developed, simulated, and experimentally measured to validate this new technique.

The investigation of the PA linearization technique depends on the PA class and operating condition, and the behavior of the PA, which requires one to model the power amplifier. The general model of the PA is the Volterra series (Schetzen, 1980), which includes the memory effects, especially the ones that cause dynamic AM-AM and AM-PM (Maas, 1997). A main drawback of the Volterra series is the high complexity that is needed to extract the coefficients, especially when the number of coefficients is increasing.

The predistortion technique used to obtain the inverse of the Volterra model is implemented by finding the inverse of the coefficients (Schetzen, 1980), which is complicated. The important point here is that deriving the exact inverse of a Volterra series is very complicated and the nth-order inverse is only an approximation. Due to the above-mentioned reason, special cases of the Volterra model are applied to model the nonlinear memory effects in the PA. The Wiener model proposed by Clark et al. (1998) is one of these special cases. The advantage of the Wiener model is the possibility for the predistorter model to be exactly inverse of the PA. Another special case for the Volterra model and the most commonly used method is the memory polynomial technique proposed by Kim (2001). Because in the Volterra series by increasing the nonlinearity order, extracting the coefficients becomes very complicated. Hence, the memory polynomial that only considers the diagonal terms of the Volterra series coefficients is the most promising technique, however, it is still not an exact compensation model. Another example of the predistortion technique is the parallel Wiener model proposed by Eun (1997).

There is not a study in which the PA model is the optimized one and then based on that, the predistortion techniques can be determined; this is due to several parameters like bandwidth, distortion, and complexity. In this chapter, predistortion linearization is the main topic; accurate PA modeling is only a secondary concern. For a nonlinear PA with memory, its inverse must also be a nonlinear system with memory. It has been shown that a memory polynomial is a good model for the PA by Kim (2001), so it is considered in this book for modeling the PA with memory.

In memoryless PAs, the nonlinearity can be compensated by applying a simple adaptation by taking feedback from the PA output. It should be noted that the AM-AM and AM-PM characteristics of the PA are static and hence by using a linear convergence technique the PA will be linearized. These characteristics are in modeling static functions extracted at a given temperature and DC bias. However, power amplifier memory effects are actually due to thermal effects or long-term memory effects. When intermodulation distortion (IMD) is measured to characterize the PA nonlinear behavior, asymmetries occur in the lower and upper sidebands (Carvalho, 2002; Cripps, 1999) and IMD magnitude variation that is related to the envelope of the signal are often observed (Ku and McKinley, 2002). This means that the AM-AM and AM-PM functions are not constant and vary depending on the past input signals. This changing behavior of the PA, called memory effects, can be modeled using the Volterra series (Carvalho, 2002; Cripps, 1999; Vuolevi et al., 2001). In Cripps (1999), the asymmetric behavior due to the memory effects are analyzed and the one called short term which makes the AM-AM and AM-PM functions scatter around their linear curve was investigated. In Carvalho (2002), a study of the reasons for IMD asymmetry is carried out by analyzing nonlinear circuits in small and large signal regions. Vuolevi et al. (2001) divided the memory effects into electrical and thermal memory effects, and described the reason for asymmetric IMD to the thermal memory effects.

The DP technique is chosen for PA linearization because of its cost effectiveness, high flexibility, and high ACLR reduction as compared to other linearization techniques (Cavers, 1990; Ding et al., 2004). However, most of the existing predistortion designs consider the PA as a memoryless device (Cavers, 1990). For wideband or high power applications, the PA has memory effects, especially the dynamic AM-AM and AM-PM functions, but memoryless predistorters can only compensate a limited amount of reduction in ACLR. There are several techniques that can compensate for the memory effects of the PA (Bondar, 2009; Boumaiza, 2003; Ding et al., 2004; Franco, 2004; Woo, 2007; Zhe, 2008). Until now, all these techniques were based on the Volterra series. The most important one is the memory polynomial predistorter, which is used in this book. The memory polynomial technique can overcome both the nonlinear distortions and the electrical memory effects that cause dynamic AM-AM and AM-PM (Ding et al., 2004). However, one of the important disadvantages of the memory polynomial is that it cannot compensate for all the memory effects due to its limitation to extract the coefficients of the PA. The ideal performance of DP is based on robust predistorters that can completely compensate for the nonlinearities of the power amplifier. In reality, other impacts called analog imperfections deteriorate DP performance, which are introduced by the analog components at the transmitter, specifically analog filters and quadrature modulators. There has been some research in field programmable gate array (FPGA) implementation (Mato, 2009). Mato (2009) implemented the DP technique in the FPGA within the hardware resources of the FPGA; however, one of the the main problems with the technique is that it does not compensate the dynamic memory effects of the power amplifier. The final design can be applied with various PAs and also in various communication systems such as a Base Transceiver Station (BTS), mobile handsets, Wireless Local Area Network (WLAN), satellite communication, Digital Video Broadcasting (DVB), and digital radio.

6.2 Power Amplifier Linearization Techniques

In this section, different linearization techniques are introduced and then DP is selected as the best one to use. The advantages and disadvantages of DP as compared to other techniques are presented. The most important linearization methods reported in the literature can be classified as feedback, linear amplification with nonlinear components (LINC), feedforward, predistortion, and DP.

6.2.1 The Feedback Linearization Technique

The simplest way of reducing amplifier distortions is by the feedback method (Patel, 2004). The use of negative feedback illustrates around an amplifier with the effect of distortion $n(t)$. L and H denote the gain of the amplifier and the feedback attenuation, respectively. The output, feedback, and the error can be expressed as (Patel, 2004)

$$Output: y(t) = L.E(t) + n(t)$$

$$Feedback: g(t) = y(t)/H \tag{6.1}$$

$$Error = x(t) - g(t)$$

Therefore,

$$y(t) = H(L.x(t) + n(t))/(L + H) \tag{6.2}$$

If the amplifier gain is much greater than the feedback ratio $L \gg H$, then $H + L$ approximates to L. So the output is (Patel, 2004)

$$y(t) = H.x(t) + (H.n(t))/L \tag{6.3}$$

Therefore, the distortion produced by the amplifier is reduced by a factor H/L. The disadvantage of this approach is that the improvement for reducing the distortion is at the expense of the gain of the power amplifier and also the feedback method loses bandwidth because of a long, high delay to find the predistortion function.

6.2.2 Linear Amplification with Nonlinear Components

Linear amplification with nonlinear components (LINC) is different from the other techniques of linearization for PAs because there is no feedback from the output of the PA. The PA can be highly nonlinear. The theory of operation is that baseband processing has a gain and phase modulated input signal, and generates two wideband modulated signals. These signals are then upconverted with two well-matched nonlinear amplifier chains and the amplified signals are then summed. The complex signal should be generated so that all undesired out-of-band components are in exact antiphase in the two amplifier chains and cancel at the output, and the wanted components are in phase.

The generation of the signals $S_1(t)$ and $S_2(t)$ should be accurate. The linearity performance of the technique is determined by the gain and phase match between the two amplifiers (Hetzel, 1991; Sundstrom, 1996).

The input signal $S(t)$ can be expressed as

$$S(t) = r(t).e^{jq(t)}; \ 0 < r(t) < r_{max} \tag{6.4}$$

This signal can be split into two signals, $S_1(t)$ and $S_2(t)$, with the modulated phase and constant amplitudes as shown in Patel (2004). This gives

$$S_1(t) = S(t) - e(t); S_2(t) = S(t) + e(t); \ and \ |S_1(t)| = |S_2(t)| = r_{max} \tag{6.5}$$

where $e(t)$ is in quadrature to the source signals $S_1(t)$ and $S_2(t)$,

$$e(t) = j.S(t).\sqrt{\left(\frac{r_{max}^2}{|S(t)|^2} - 1\right)} \tag{6.6}$$

The output is given by

$$S_{out}(t) = G.2S(t) \tag{6.7}$$

The quadrature signal $e(t)$ is added to one leg of the forward loop and subtracted from the other leg of the forward loop to provide a constant envelope signal.

The main disadvantage of this approach is that this technique generates two constant envelope signals, which is complicated, and high power combined with low loss and high isolation is very difficult to achieve.

6.2.3 Feedforward Linearizers

The feedforward linearizer was first proposed by Harold S. Black (1977) when he was working for the Western Electric Company and was trying to improve Bell System's new telephone system. Feedforward is known to be a suitable linearizer for operating with wideband signals. In the two-tone test, the input signal is split equally and fed to the upper and lower paths, respectively. The signal in the upper path is then amplified by the PA. The output signal of the PA shows the IMD products.

An attenuated sample of the PA output signal is inserted into the lower path and then subtracted with a delayed version of the input signal. Therefore, assuming an ideal match between the lower path delay and the delay introduced by the PA, the resulting error signal includes only the IMD products caused by the PA. Then, the error signal is amplified by the error amplifier and inserted into the upper path (the error is subtracted). The PA output signal in the upper path should be delayed by an amount equal to the delay of the error amplifier. So then, an amplified version of the desired two-tone signal will be at the output of the error that is subtracted. The main disadvantage of this technique is that it is not stable. Since it is an open loop, this makes it too sensible to delay the mismatches and tolerances of the components. These mismatches may produce imbalances between the different branches of the linearizer structure. If these imbalances are not compensated, the imperfect cancellation in both paths can cause a reduction in performance (Pothecary, 1999). It can also cause significant power efficiency degradation even when the linearity levels are maintained (Gilabert et al., 2004). In addition, the same open-loop nature of the linearizer makes it too sensitive to nonlinearities and losses that are produced by the loop components.

6.2.4 Predistortion Linearizers

The basic principle of predistortion is to create an inverse function for the power amplifier. So, the AM-AM characteristic of the predistortion block should be inverse and the AM-PM should be negative for the power amplifier AM-AM and AM-PM, respectively. The result of the predistortion series with a power amplifier will be a linear gain at the output. With this simple technique, the efficiency of the system will increase significantly and the out-of-band distortion will be reduced.

There are different ways to accomplish predistortion. It can be implemented in the baseband (BB), which is discussed in this book and which can also be found in intermediate frequency (IF) and radio frequency (RF).

It should be noted that most current predistorters use feedback to make an adaptive predistorter compensate for memory effects; mainly long-term memory can be compensated in this way. But the other type of memory effects, short term, cannot be compensated with a simple adaptation. Predistortion can be applied at RF, IF, or BB. The advantage of predistortion at IF and BB is that they are independent on the final frequency band of operation, in addition to being more robust in terms of environmental parameters. Also, the cost of ADCs (analog-to-digital converters) and DACs (digital-to-analog converters) decreases at low frequency operation. The main drawback regarding predistortion compensation at the IF or BB is that there are more linearity requirements since the upconverters can introduce additional distortion. However, upconverters can be avoided at the IF by using software radio techniques.

6.2.5 Digital Predistortion

Digital predistortion (DP) is a technique that can potentially meet the requirements of any communication system as it is the most cost effective and most efficient way for linearity of the power amplifier. The reason that the cost of this technique is less than the feedforward technique is that digital predistortion is implemented in baseband but feedforward needs an extra PA and is implanted in RF. In baseband, the cost of implementation is much less than in RF. Figure 6.1 shows a block diagram for a DP system. According to Figure 6.1, the predistortion

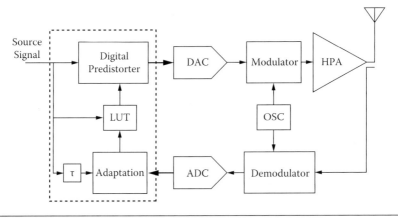

Figure 6.1 Block diagram of digital predistortion. (From Kenington, P. B. 2000. *High-Linearity RF Amplifier Design,* Norwood, MA: Artech House. With permission.)

function contains a lookup table (LUT), complex gain multiplier, and most important, an adaptation block. The main concentration here is on the adaptation part, which will be used for compensating the memory effects of the PA. The details of these parts are explained next.

An LUT is a data structure, usually an array, used to store a predistortion function. The LUT uses RAM to store outputs for corresponding inputs and can either be a two-dimensional (2D) table or a one-dimensional (1D) table. An n-bit LUT can be implemented with a multiplexer where select lines are the inputs of the LUT and whose outputs are constants. An n-bit LUT can encode any n-input Boolean logic function just by modeling such functions as truth tables. This is an efficient way of encoding Boolean logic functions.

Most adaptive DP blocks are designed to be capable of adapting to the slow variation of PA characteristics, however, the fast variation is not taken into consideration. Adaptation to the fast variation of PA characteristics, which are considered to be short-term memory effects, with conventional predistortion schemes these can degrade the performance or even make the system unstable. The LUT technique requires an ADC to index the two tables. As an example, assume the 2D LUT is in a polar format and one table includes the scale (S) and another one includes rotation (R) of the gain of predistortion that is being multiplied in the input signal to produce the desired input signal to pass through the power amplifier. Because of the large number of table entries that need to be optimized, adaptation is typically performed at the baseband. The error signal, which is the result of a subtraction operation between the output and input signals, is used as an update parameter for the LUT entries. This method requires a sample-by-sample comparison between the input and output, increasing the complexity but improving the overall suppression. Otherwise, the table entries could be optimized based on adjusting the table entries to conform to a higher-order function, reducing the complexity but with a degraded suppression performance (Cavers, 1990). The amplitude of the input signal is used to index the LUT.

The complex gain multiplier is the other block of the predistortion. The complex input signal is multiplied by the complex LUT values in the complex multiplier block.

In order to adapt the power amplifier to changes in its characteristics, adaptation is crucial and needs to be performed based on the error from the input and output of the power amplifier. The most common techniques to adapt power amplifier behavior are discussed next.

The secant method for adaptation uses a root finding problem, which is the error generating from the difference between the input and the normalized output. The root has a relationship with the table entries and they are adjusted using the conventional secant method (Sundstrom, 1996).

However, the linear convergence method is faster and less complex. This approach uses a normalized error generating from the output and input samples, which update the LUT values iteratively. The scaling parameter controls the convergence rate (Cavers, 1990).

In Table 6.1, the comparison of the different linearization techniques is presented in terms of ACLR correction, bandwidth, efficiency, flexibility, and cost. It can be seen that the DP technique is more applicable than the other techniques. As discussed earlier, DP will be implemented in the FPGA as the adaptation is based on digital data, and the output samples from the PA should be downconverted in the baseband, which is the main reason that this technique costs less compared to the feedforward technique. And also as shown in Table 6.1, the DP technique is more flexible than other techniques because the implementation of this technique in the baseband causes this technique to have more flexibility in terms of choosing the adaptation parameters and the ability to modify it based on the requirements.

Table 6.1   Comparison of Linearization Methods

TECHNIQUE	CORRECTION (ACPR)	BANDWIDTH	EFFICIENCY	FLEXIBILITY	COST
Addition of a nonlinear component	3–5 dB	15–25 MHz	5%–8%	Low	Low
Cartesian loop feedback	15–20 dB	<5 MHz	10%–12%	Medium	Medium
Feedforward linearization	30–35 dB	>100 MHz	6%–10%	Medium	High
Digital predistortion linearization	20–30 dB	50–100 MHz	12%–14%	High	Medium

There are two important methods that are used for obtaining the novel predistortion method, which will be explained next. Each of these techniques has advantages and disadvantages.

6.2.6 Memory Polynomial Predistortion

A memory polynomial predistortion is a special case of the Volterra series in which only the diagonal kernels are considered. In this section, the memory polynomial predistorter (Ding et al., 2004; Kim, 2001; Ma et al., 2001) is explained and is used in simulations in Chapter 5 for comparison with our method. The predistorter technique is based on the indirect learning architecture without the need of the power amplifier model (Figure 6.2). Compared to the Hammerstein predistorter, the memory polynomial predistorter has more terms. However, its parameters can be easily estimated by using least squares in the training branch. The memory polynomial predistorter can be described by

$$z(n) = \sum_{\substack{k=1 \\ odd}}^{K} \sum_{q=0}^{Q} b_{kq} \, y(n-q) \left| y(n-q) \right|^{k-1} \tag{6.8}$$

where $y(n)$ and $z(n)$ are, respectively, the input and output of the predistorter in the training branch; b_{kq} is the coefficient of the predistorter; K is the order of nonlinearity, which here the odd orders are only considered; and Q is the memory length, which defines the amount of the memory that can be modeled. Since the model in Equation (6.8) is linear with respect to its coefficients, the predistorter coefficients b_{kq} can be obtained directly by least squares.

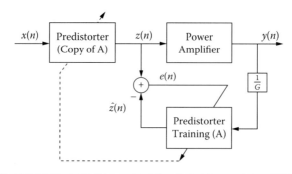

Figure 6.2 The indirect learning architecture for the predistorter. (From Ding, L. et al. 2004. *IEEE Transactions on Communications* 52(1):159–165. With permission.)

By defining a new sequence

$$u_{kq}(n) = \frac{y(n-q)}{G} \left| \frac{y(n-q)}{G} \right|^{k-1} \tag{6.9}$$

at convergence, we should have

$$z = Ua \tag{6.10}$$

where

$$z = [z(0),...z(N-1)]^T$$

$$U = [u_{10},...u_{K0},...,u_{1Q},...,u_{KQ}]$$

$$u_{KQ} = [u_{kq}(0),...,u_{kq}(N-1)]^T \tag{6.11}$$

$$a = [a_{10},...,a_{K0},...,a_{1Q},...,a_{KQ}]^T$$

The least-squares solution is

$$a = (U^H U)^{-1} U^H z \tag{6.12}$$

where $(.)^H$ denotes a complex conjugate transpose.

Figure 6.3 shows the flowchart for the memory polynomial predistortion technique.

6.2.7 Complex Gain Predistortion

In the complex gain predistortion technique, the adaptation is based on the linear convergence method, which is a faster and simpler method to implement. This approach uses a scaled version of the resulting error

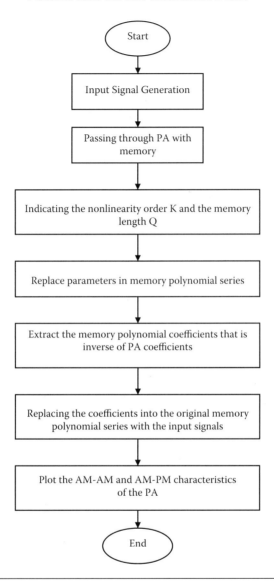

Figure 6.3   Flowchart for the memory polynomial predistortion.

signal to adjust the LUT entries iteratively. The scaling parameter controls the rate of convergence. And indexing an LUT is done by calculating the absolute of the input signal, which is a complex number in simulation by programming in MATLAB. In Figure 6.4, it is shown that between the LUT and the complex gain adjuster block, data needs to be converted from the polar format or R and S to the rectangular format or I and Q to be in the same format as the input signal.

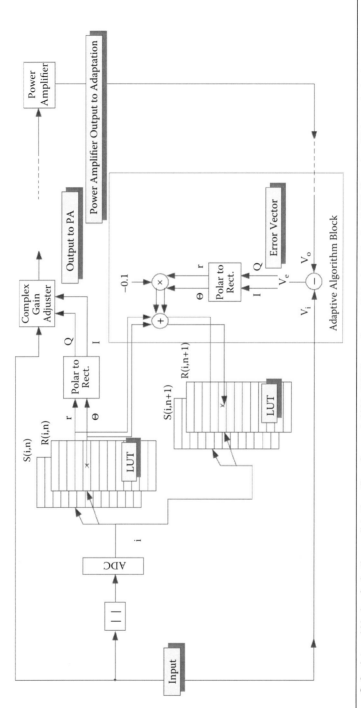

Figure 6.4 Internal blocks of adaptive predistortion. (From Cavers, J. 1990. *IEEE Transactions on Vehicular Technology* 39(4), November. With permission.)

The equations that are being used to adjust the LUT data are

$$S_{(i,n+1)} = S_{(i,n)} - 0.1 \times |V_e|$$

$$R_{(i,n+1)} = R_{(i,n)} - 0.1 \times \angle V_e$$

(6.13)

The coefficient of -0.1 is the scale parameter and V_e is the error signal, calculated from the output of the power amplifier and the input, which is shown by a minus sign in the adaptation algorithm block. The error voltage from the output of the power amplifier and input signal can be expressed as

$$V_e = |V_e| \angle V_e = V_o - V_i$$

(6.14)

By calculating the error vector and using the updated equations, the content of the LUT will be updated and after finishing the work in the simulation program, the optimum amount for the LUT will be found, and so the predistorted output of the power amplifier will be made.

Memory effects can degrade the performance of the DP just like linear distortions, so these effects must be reduced. Memory effects are more difficult to deal with than linear distortions because they are produced by several factors such as aging, temperature variation, antenna mismatch, and changing the behavior of lumped elements.

Table 6.2 reviews the weaknesses and strengths of the most relevant publications on the effects of memory in digital predistortion for compensating nonlinearity in power amplifiers. Many papers describe methods to measure memory effects in RF power amplifiers (Bosch, 1989; Ku, 2003; Vuolevi et al., 2001). These methods require the use of test equipment, so are not practical in a digital predistorter implementation. Other papers (Carvalho, 2002; Jeckeln et al., 2003) propose memory effect models, which are again not practical in a digital predistorter implementation, although they do provide some research on the causes of memory effects. In recent years, there have been some studies to measure and correct for memory effects in RF power amplifiers (Ding et al., 2004; Franco, 2004; Woo et al., 2007). Not all major memory effects are considered in the reviewed research and only two of the papers show experimental results with actual memory effect corrections (Woo, 2007; Zhe, 2008). Most of the previous studies in the area of memory effects are used with the Volterra series. Ding et al. (2004) applied the memory polynomial

Table 6.2 Review of the Literature on Memory Effects in Digital Predistortion

REFERENCES	TOPICS	WEAKNESSES	STRENGTHS
Ku, 2003	Measurement of memory effects	Does not show linearization results	Propose an accurate polynomial model with delay taps
Franco, 2004	Minimization of memory effects	Provides few details on the digital solution	Identifies sources of memory effects and provides analog and digital methods for correction
Vuolevi, 2003	Measurement of memory effects	Methods are complicated and impractical for a digital linearizer	Analyzes thermal and electrical memory effects
Ding et al., 2004	Digital predistorter with memory polynomials	Just simulation results; implementation needs a lot of hardware resources	Compensates for dynamic memory effects
Carvalho, 2002	Analysis of asymmetry in distortion sidebands	Simulation is on the device not the technique	Performance is improved in terms of table size and quantization effects
Zhe, 2008	Memory effect compensation based on the Volterra series	Highly complex	It includes more memory coefficients
Woo, 2007	Dynamic memory effects	Cannot compensate for all the memory contents	Investigates the method that model the PA with memory
Morgan et al., 2006	Memory effect techniques	Does not work significantly on any technique	Compares all the techniques together

technique, which is a special case of the Volterra series, and is capable of compensating the dynamic memory effects produced by the power amplifier. However, the main drawback with this technique and other proposed methods was that by increasing the number of coefficients and memory length of the predistorter, extracting these coefficients became difficult especially in implementation.

Some factors that can be used to evaluate the performance of a DP technique include linear distortion, and memory effects. Until now, most of the studies have been carried out in a simulation environment, however, the practical implementations have not been addressed very well. There are also some studies on analog imperfections as a result of a DAC and mixer. They generate DC offset leads to power efficiency degradation, and phase and gain imbalances. A DP technique has to be able to compensate for these effects too. The design of the predistortion technique depends on several other factors such as bandwidth requirements, power efficiency, adjacent channel power ratio (ACPR), and complexity. There are trade-offs between these factors when selecting and designing a predistortion technique. Very few papers have performed analytical and experimental results by utilizing all sources of nonlinearity and they permit the measurement of nonlinear distortion at certain signal envelope frequencies (Bondar, 2009; Morgan et al., 2006).

Advantages and Disadvantages of Linearization Techniques

Since an optimal linearization technique does not exist, each linearization technique can be considered for some particular applications, so there is always a trade-off, such as:

- Linearity requirements
- Complexity that adds linearizers with regard to hardware resources
- Signal bandwidth of the linearizer
- Technology of the linearization based on analog circuitry or digital processing devices
- Power efficiency requirements, trade-off between linearity and efficiency

6.2.8 The Digital Predistortion Linearization Method

The DP circuit operates on baseband signals, whereas in other cases, the predistortion circuit is configured to operate on intermediate frequency (IF) signals. The nonlinear correction may be applied to signals using any suitable form of digital signal processing, including both real and complex domain (I/Q or polar) processing. Preferably, the nonlinear correction is applied by using a parallel processing architecture, whereby two or more samples are processed simultaneously, in order to accommodate the high sample rate of the expanded bandwidth.

It is also important when studying the predistortion method, that the predistortion attempts to add third- and fifth-order intermodulation products to the input signals which cancel out the third- and fifth-order intermodulation products added by the PA. Thus, the bandwidth of the predistorted signal must be three times greater than the bandwidth of the input signals to be able to represent up to fifth-order intermodulation products. In the real world, the predistorted signals are fed into a DAC and then low-pass filtered at the Nyquist rate (half the input sample rate). The predistorted signal must have a sample rate of at least six times that of the original input signals. Therefore, in simulations the input signals are interpolated by a factor of six before being fed into the predistorter.

Figure 6.5 shows a block diagram for adaptive digital predistortion. A fully adaptive DP system requires the addition of a predistortion circuit consisting of a digital predistorter and LUT to the transmission path in addition to a feedback path consisting of a demodulator, ADC, and adaptation circuit for updating the LUT. The block diagram assumes that all components of the system except the predistorter and high power amplifier (HPA) have a linear response and hence can be ignored in the analysis. Also, these effects (ADC, DAC, upconverter, and downconverter) are ignored here.

The predistorter is equivalent to a nonlinear circuit with a gain expansion response that is inverse for the power amplifier gain compression AM-AM and a phase rotation that is negative to the power amplifier phase rotation AM-PM. In Figure 6.5, $X(n) = I + jQ$ is the quadrature modulated input signal and $V_f(n)$ is the quadrature demodulated feedback signal. These signals are synchronously sampled,

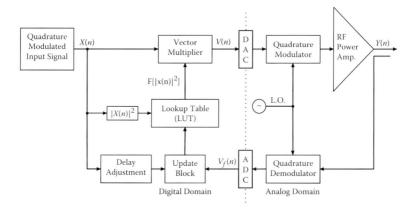

Figure 6.5 Adaptive digital predistortion block. (From Kenington, P. B. 2000. *High-Linearity RF Amplifier Design,* Norwood, MA: Artech House. With permission.)

and their values are used to generate a predistortion vector function $F[|x(n)|^2]$, which is stored in a polar or rectangular form in an LUT. The input signal is $x(n)$ predistorted according to $F[|x(n)|^2]$, so that the predistorted signal $V(n)$ produces the linearized output from the RF amplifier. Here, the LUT is 10 bits and the absolute of the input signal is used for addressing it.

One of the main objectives here is to study the electrical memory effects that cause dynamic AM-AM and AM-PM. Previous studies were restricted to the calculation of the coefficients of the power amplifier. This way needs a lot of computation and therefore takes a lot of processor time, and can never be implemented when the number of coefficients increases. The technique that is proposed here does not have that drawback. It even claims that it can compensate the dynamic memory effects in wideband applications. This method is discussed in detail in the following section.

6.2.9 Complex Gain Memory Predistortion

Figure 6.6 shows the block diagram of predistortion cascades with the power amplifier denoted $P[|x(n)|^2]$ and $S[|r(n)|^2]$, respectively. It should be noted that $P[|x(n)|^2]$ and $S[|r(n)|^2]$ contain a complex gain representation of predistortion and the power amplifier functions, respectively.

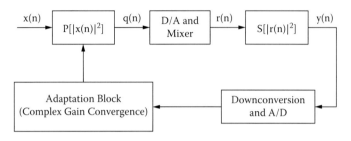

Figure 6.6 The block diagram for the predistortion series with the power amplifier.

In Ding et al. (2004), the power amplifier model consists of memory effects, mainly the electrical memory effects or short-term effects that cause the power amplifier characteristics to spread, which can be expressed by

$$y(n) = \sum_{\substack{k=1 \\ Odd}}^{K} \sum_{m=0}^{M} c_{km} r(n-m) \left| r(n-m) \right|^{2(k-1)} \tag{6.15}$$

where $r(n)$ and $y(n)$ are the power amplifier input signal and the output complex signal, respectively, and K and M are the nonlinearity order and the memory length, respectively.

The power amplifier input signal can be described as follows

$$r(n) = I(t)\cos(\omega t) - Q(t)\sin(\omega t) \tag{6.16}$$

with $I(t)$ and $Q(t)$, the baseband modulation signals.

It should be noted that Equation (6.1) only considers the odd-order nonlinear terms, this is because of the fact that the even terms are filtered out by a band-pass filter that causes intermodulation distortion. The modified input signal of the power amplifier $r(n)$ can be expressed by

$$r(n) = x(n)P[\left| x(n) \right|^2] \tag{6.17}$$

where $x(n)$ is the input data stream and $P[\left| x(n) \right|^2]$ is the predistortion function. After some simplification, we have

$$y(n) = \sum_{m=0}^{M} r(n-m) \sum_{\substack{k=1 \\ Odd}}^{K} c_{km} \left| r(n-m) \right|^{2(k-1)} \tag{6.18}$$

where $S_m(|r(n-m)|^2)$ can be given by

$$S_m[|r(n-m)|^2] = \sum_{\substack{k=1 \\ Odd}}^{K} c_{km} |r(n-m)|^{2(k-1)} \qquad (6.19)$$

and,

$$y(n) = \sum_{m=0}^{M} r(n-m) S_m[|r(n-m)|^2] = r(n) S_0[|r(n)|^2]$$

$$+ r(n-1) S_1[|r(n-1)|^2] + \dots \qquad (6.20)$$

From the preceding equations it can be shown that the power amplifier model shown by the memory polynomial is not the only way to represent the memory of the power amplifier, and in fact the other way to model the memory effects according to Equation (6.20) is based on the complex gain coefficient modeling. Here, it is shown that the proposed scheme represents more memory content, and in addition, the complexity to extract the coefficients of the memory polynomial is resolved.

The power amplifier in ideal conditions has the following linear characteristics:

$$y(n) = Sx(n) \qquad (6.21)$$

where S denotes the power amplifier linear gain.

In another way, Equation (6.20) explains that by having a distortionless transmission through a power amplifier we require that the exact input signal shape be produced at the output, although its amplitude may be different and it may be delayed in time. Therefore, if $x(n)$ is the input signal (it is assumed that the power amplifier is ideal and hence there is no predistortion block, thus $r(n) = x(n)$, and the required output is

$$y(n) = Sx(n - n_d) \qquad (6.22)$$

where n_d is the delay. Taking the Fourier transform of both sides of Equation (6.22), we get

$$Y(\Omega) = Se^{-j\Omega n_d} X(\Omega) \qquad (6.23)$$

Thus, it can be concluded that for having a distortionless transmission, the system must have

$$H(\Omega) = |H(\Omega)| e^{j\theta(\Omega)} = Se^{-j\Omega n_d} \qquad (6.24)$$

Hence

$$|H(\Omega)| = S$$
$$\theta(\Omega) = -j\Omega n_d \qquad (6.25)$$

That is, the amplitude of $H(\Omega)$ must be constant over the entire frequency range, and the phase of $H(\Omega)$ must be linear with the frequency.

Hence, from Equations (6.20) and (6.21), we have:

$$y(n) = \sum_{m=0}^{M} r(n-m)S_m[|r(n)|^2] = Sx(n) \qquad (6.26)$$

Now the main goal is to extract the optimum predistortion function from the complex gain criteria in Equation (6.26). Following some simplification, the optimum predistortion function is derived from the following iterative expression:

$$P_{i+1}[|x(n)|^2] = P_i[|x(n)|^2] - \frac{P_i[|x(n)|^2]}{r(n)S_0[|r(n)|^2]} V_{error}(n) \qquad (6.27)$$

where

$$V_{error}(n) = y(n) - Sx(n) \qquad (6.28)$$

Equation (6.27) is the predistortion function while the power amplifier has no memory, or the memory length is zero. In fact, the power amplifier exhibits memory effects, which requires a more general formula to consider those effects (see Figure 6.7).

Now if we assume that the power amplifier memory term sets to 1, in other words $M = 1$, the predistortion function is given by

$$P(|x(n)|^2) = \frac{S}{S_0[|r(n)|^2]} - \frac{r(n-1)S_1[|r(n-1)|^2]}{x(n)S_0[|r(n)|^2]} \qquad (6.29)$$

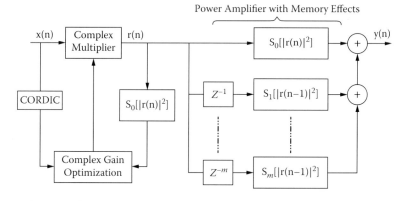

Figure 6.7 The predistortion block diagram with memory effects.

By applying the iterative solution to Equation (6.29), the following function is derived:

$$P_{i+1}\left(\left|x(n)\right|^2\right) = P_i\left[\left|x(n)\right|^2\right] - \frac{P_i\left[\left|x(n)\right|^2\right]}{r(n)S_0\left[\left|r(n)\right|^2\right]}V_{error}(n)$$

$$+ \frac{P_i\left[\left|x(n)\right|^2\right]r(n-1)S_1\left[\left|r(n-1)\right|^2\right]}{r(n)S_0\left(\left|r(n)\right|^2\right)} \qquad (6.30)$$

$$- \frac{r(n-1)S_1\left[\left|r(n-1)\right|^2\right]}{x(n)S_0\left[\left|r(n)\right|^2\right]}$$

This equation can be simplified as follows:

$$P_{i+1}\left(\left|x(n)\right|^2\right) = P_i\left[\left|x(n)\right|^2\right] - \frac{P_i\left[\left|x(n)\right|^2\right]}{r(n)S_0\left[\left|r(n)\right|^2\right]}V_{error}(n)$$

$$(6.31)$$

$$+ \frac{r(n-1)S_1\left[\left|r(n-1)\right|^2\right]}{S_0\left[\left|r(n)\right|^2\right]}\left(\frac{P_i\left[\left|x(n)\right|^2\right]}{r(n)} - \frac{1}{x(n)}\right)$$

From Equation (6.31) it can be observed that the second fraction becomes zero due to the fact that $v(n) = x(n)P_i\left[\left|x(n)\right|^2\right]$

$$P_{i+1}\left(\left|x(n)\right|^2\right) = P_i\left[\left|x(n)\right|^2\right] - \frac{P_i\left[\left|x(n)\right|^2\right]}{r(n)S_0\left[\left|r(n)\right|^2\right]}V_{error}(n) \qquad (6.32)$$

It is obvious that the predistortion function in Equation (6.32) is the same as the one in Equation (6.29), except that the error vector was obtained from the difference of the output and the normalized input. This is because the power amplifier in Equation (6.26) contains more memory terms. The power amplifier memory terms can be extended to include more memory terms. As shown in Ding et al. (2004), the memory polynomial is built on a special case of the Volterra series in which only the diagonal terms in the coefficient matrix are considered; however, this technique is unable to compensate for the large memory effects, especially when the memory terms are $M > 3$. However, the CGC scheme can compensate for more memory terms. The main reason for this is that compared to the memory polynomial where the complexity rises significantly when the memory length increases is that in this scheme the predistortion function is related to the complex gain function.

The important parameter in Equation (6.27) is the ratio of $P[|x(n)|^2]$ over $r(n)S_0[|r(n)|^2]$, which is called the gain factor. According to Liu et al. (2013) and Morgan et al. (2006), the gain factor is a constant between 0 and 1. A higher value has faster convergence but lower performance, and a lower value of the gain factor yields higher performance but slower convergence. In the complex gain convergence (CGC) scheme, the α parameter represents the gain factor, which is not constant due to the memory effect modeling in the CGC scheme. To get control of the complex gain convergence as shown in Equation (6.27), the general form of the CGC scheme can be expressed by

$$P_{i+1}(|x(n)|^2) = P_i[|x(n)|^2] - \alpha \frac{P_i[|x(n)|^2]}{r(n)S_0[|r(n)|^2]} V_{error}(n) \qquad (6.33)$$

where α is a constant and its value has to be between 0 and 1. To verify the feasibility of the proposed scheme in actual systems, the analysis of the complexity of the gain factor that includes division and multiplication needs to be investigated.

To devise the CGC scheme with less complexity, the simplification is performed as follows:

$$\frac{P[|x(n)|^2]}{r(n)S_0[|r(n)|^2]} = \frac{P[|x(n)|^2]}{x(n)P[|x(n)|^2]S_0[|r(n)|^2]} = \frac{1}{x(n)S_0[|r(n)|^2]} \qquad (6.34)$$

It is assumed here that the complex gain function of predistortion $P[|x(n)|^2]$ is 1, hence:

$$\frac{1}{x(n)S_0[|v(n)|^2]} = \frac{1}{x(n)S_0[|x(n)|^2]} = \frac{1}{y_0(n)} \tag{6.35}$$

where $y_0(n)$ is the output of the power amplifier with no memory effects. As the main goal is to have a linear system:

$$y_0(n) = Sx(n) \tag{6.36}$$

The convergence criteria in Equation (6.33) is achieved with two or three iterations, which shows faster convergence, better performance, and less complexity compared with the indirect learning architecture method in Ding et al. (2004). The other metric that can be used to evaluate the performance of the proposed predistortion scheme is error vector magnitude (EVM) defined by

$$EVM = \frac{rms(|V_{error}(n)|)}{rms(|(x(n)|)} \tag{6.37}$$

Advantages and Disadvantages of the Complex Gain Memory Predistortion Method

The complex gain memory predistortion (CGMP) method has many benefits compared to the other methods discussed until now. (See Figure 6.8.) One of these is that it does not need to calculate the coefficients of the Volterra series, which is very complicated to extract. The other important feature is that this method can compensate for all the memory effects that the power amplifier contains, especially the ones that cause dynamic AM-AM and AM-PM.

The only drawback of this method is that to calculate the inverse of the input data, which is after finding it and multiplying with the error vector, then the LUT content could be updated. So in implementation, the main concern is the division part, which is the main issue in implementation and will be discussed in detail in the next chapter. Equation (6.33) with two or three iterations is convergence and it is shown that as compared with the indirect learning architecture method in Ding et al. (2004), the efficiency improves more (around 3%) and is less complex. It will be shown that if Case 1 (Equation 6.33) is applied in simulations, the speed of convergence is more than in Case 2 (Equation 6.36), which requires more iteration for convergence. The only time-consuming part for implementing this method is the calculation of the gain factor, which requires the inverse of the input signal.

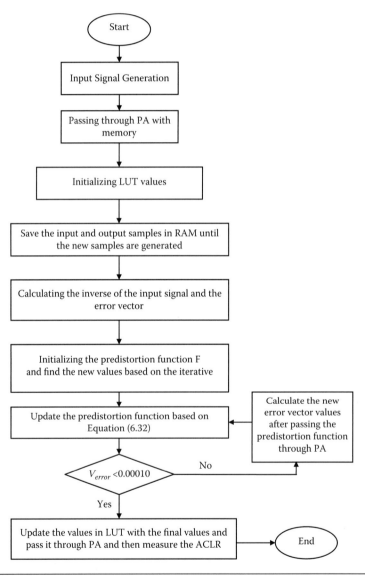

Figure 6.8 Flowchart for the complex gain memory predistortion (CGMP) method that is proposed here.

Next, the results of applying the CGMP technique and how it compares to the other memory polynomial technique are shown. But first the results of the power amplifier simulation are presented.

6.3 Simulation Results of Applying Complex Gain Memory Predistortion

In all the simulations here, the effects of analog imperfections are not considered, and also the effects of a DAC and ADC and the quantization effects are ignored (Cavers, 1997). In Appendix A, the complex baseband representation of band-pass signals is shown and it is proved that this model could be used in the simulations without the need to include the upconversion and analog parts.

Here, the presented method is successfully tested with these two types of PA models. In all the simulations, the input backoff (IBO) means that the voltage IBO is 3 dB except if it is mentioned, and the method that is applied here is Case 2 (Equation 6.36), which was discussed above.

Figure 6.9 shows the AM-AM characteristics of the PA when applying the CGMP method. In Figure 6.9a, one of the characteristics is when the predistortion has one iteration, and Figure 6.9b shows five iterations being applied. As can be seen, the IBO is 2 dB, which is very close to the saturation point, and in this figure, around this point the predistortion suffers from linearizing the PA. To overcome this impact, the gain factor parameter in Equation (6.30) can be varied, which controls the convergence rate of the iteration equation, and slowly reaches the saturation point without losing accuracy. But definitely more iterations should be applied, which increases the cost of the system. In Figure 6.9 the α parameter is considered to be 1, which is the maximum value. It is also clear that when the CGMP technique is applied, the AM-AM characteristic is forced to be linear and it can be seen that the scattering is decreased while the iteration is increased. In Figure 6.10, the AM-PM characteristics of the PA when the IBO is 2 dB is shown. According to Figure 6.10, the same problem happens when the input signal is going to increase. This becomes worse when an effort is made to reach the saturation point. The x and y axes are normalized in these figures. The spreading of samples in Figure 6.10 is is caused by the memory effects.

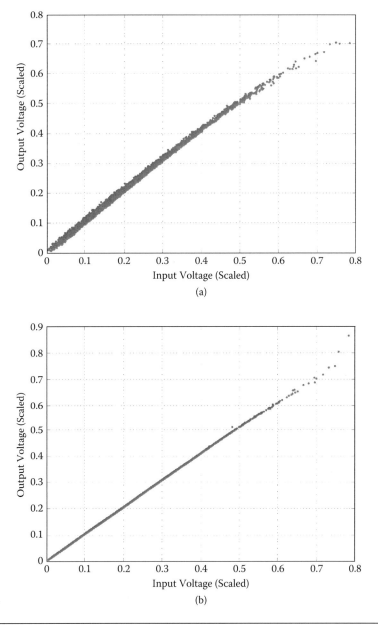

Figure 6.9 AM-AM characteristics of the PA after applying the CGMP method when the input backoff is nearly 2 dB: (a) after one iteration and (b) after five iterations.

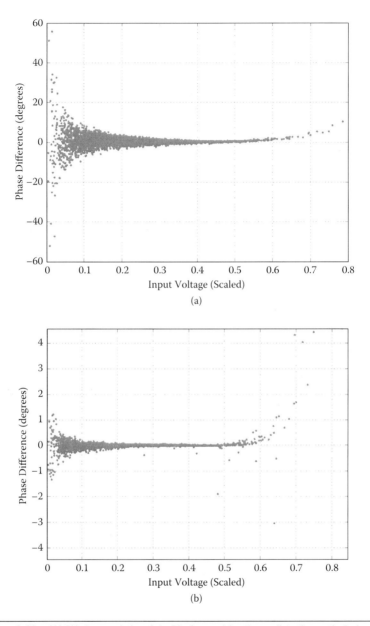

Figure 6.10 AM-PM characteristics of the PA after applying the predistortion method when the input backoff is nearly 2 dB: (a) after one iteration and (b) after five iterations.

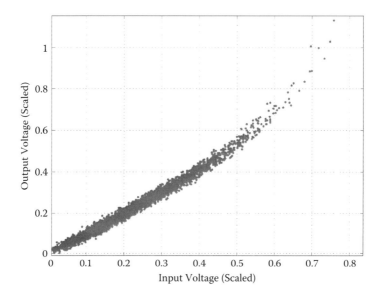

Figure 6.11   Power amplifier input after applying the predistortion function with five iterations.

Figure 6.11 shows the power amplifier input after applying the CGMP technique. It can be seen that the characteristic of this signal is the inverse of the PA and it is actually the predistortion job that should create the signal after applying it to the PA: then the linear characteristics are the result.

Figure 6.12 shows the CGMP function, which is extracted after five iterations. As shown, the complex values of the predistortion function are scattered around the gain value of 1, which is correct because the values of the input signals do not create nonlinearity and are small values, so predistortion should not have any impact on these samples. Then predistortion remains as 1 until the AM-AM characteristic of the PA is not linear anymore, and so the CGMP is applied to update the predistortion function. Actually, these values are after five iterations.

Figure 6.13 shows the error voltage of the power amplifier and input signal after five iterations. It can be seen that the error values are not in convergence when it reaches the saturation point. Also, when the input voltage is increased the error is increased too because at a higher voltage near the saturation point the adaptation formula hampers convergence and causes the error to be increased. The value of the error vector when the IBO is more than 3 dB is in the order of 0.0001, but here the IBO is 2 dB. It can be seen that the PA with predistortion

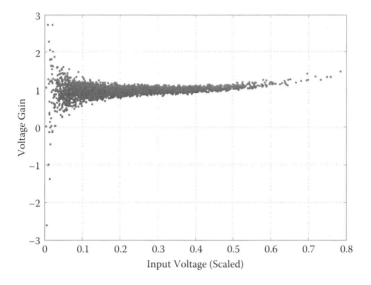

Figure 6.12   Complex gain memory predistortion function after five iterations.

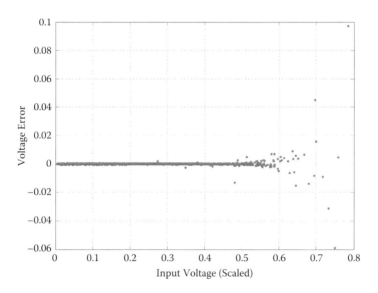

Figure 6.13   Error voltage of the power amplifier after five iterations.

is hampered and the CGMP technique is trying to decrease the error
in the regions where the input power is near the saturation point, as
shown in Figure 6.13, the CGMP is going to be nonconvergence.
It can also be seen that the value of the error vector is negative and
this is true because the AM-AM characteristic of the PA is bending,

the output is less than input voltage, and the positive points are because of the memory effects that spread in the AM-AM.

The objective of the power amplifier linearization is, first, suppression of spectral regrowth to reduce adjacent channel interference and, second, minimization of in-band distortion to improve the bit error rate (BER). Although only power spectral density (PSD) plots are discussed here, this does not mean that in-band distortion is left unchecked. Recall that in the CGMP technique, our convergence criterion requires the mean squared error between y(n) and Gx(n) to be minimized. Therefore, at convergence, the power amplifier is linearized, which automatically ensures the suppression of both in-band and out-of-band distortions. The PSD plots are shown for verification purposes. In Figure 6.14, the PSD of the power amplifier with memory is shown for different input backoffs. As seen, the backoff decreases, the nonlinearity increases, and especially near the saturation point the nonlinearity increases significantly.

For modeling the memory effects of the power amplifiers, Ku and Kenney (2003) proposed a method that is based on spar delay taps and is able to take into account all the memory effects of the power amplifier. The memory effect modeling ratio (MEMR) was used to

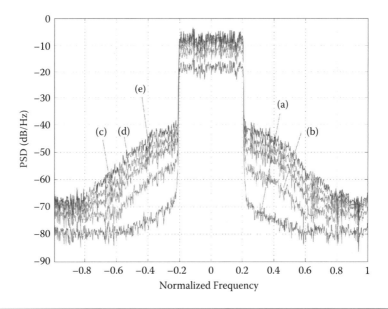

Figure 6.14 Power spectral density (PSD) of the power amplifier with memory for different IBO with WiMAX signals: (a) 6 dB IBO, (b) 5 dB IBO, (c) 3 dB IBO, (d) 2 dB IBP, and (e) 1 dB IBP.

show the amount of memory that this method can model. The power amplifier that is designed here has a MEMR = 0.45 and the one in Ku and Kenney (2003) has a MEMR = 1; these coefficients are shown in Table 6.3. Previous studies have shown that the comparison of the power amplifier with MEMR was less than 1.

In Table 6.3 the proposed method is compared with the memory polynomial method for two different power amplifiers with different memory content for a two-carrier WCDMA signal. The comparison is done with the ACLR and EVM factors.

Figure 6.15 shows the measured PSD of the two different power amplifiers with memory when a two-carrier WCDMA signal is applied. The PSD that is used in all the simulations is based on the Welch spectral estimator and the Hann window with a 1024 segment length. Figure 6.15a is for the PA having a MEMR = 1 and Figure 6.15b is the PA with a MEMR = 0.45. It can be seen that the out-of-band distortion, especially IM3 (third intermodulation distortion) and IM5, is more in the power amplifier with a MEMR = 1 because of the higher number of memory effects that are modeled in this PA. The other effect that can be seen in Figure 6.15, is the asymmetry of the distortion in which the left side and right side have different ACLRs. Ku and Kenney (2003) proposed the method for reducing such effects. The other interesting thing that it is shown in Figure 6.15 is the power amplifier with a MEMR = 1, with the order of nonlinearity (K) that is 2, only has IM3, and for the power amplifier with a MEMR = 0.45 the value is 3, and includes the IM3 and IM5.

Figure 6.16 shows the results of applying the CGMP techniques. In Figure 6.16, when applying the memory polynomial technique the amount of reduction in out-of-band distortion is on average −39.5 dB and the EVM is 0.89%, but when applying the CGMP technique after nine iterations it is almost like the input signal and these values are −53 dB and 0.71%, respectively. This means there is about a 13.5 dB improvement in ACLR as compared to the memory polynomial method, which means that an improvement of between 25 dB to 30 dB in ACLR is achieved when applying the CGMP technique.

As shown here, by increasing the iteration the amount of ACLR reduction is also increased, but it has the drawback in practical implementation to reduce the bandwidth because there is an inverse relation between the convergence time and required bandwidth.

Table 6.3 Comparison of the Two Predistortion Techniques for Different PAs with a Two-Carrier WCDMA Signal

PREDISTORTION TECHNIQUE	POWER AMPLIFIER COEFFICIENTS	MEMR	ACLR (DBC)		EVM (%)
			LEFT	RIGHT	
Memory polynomial	$a_{10} = 0.9800 - 0.300i; a_{11} = 0.06 + 0.03i; a_{12} = 0.02 + 0.08i; a_{13} = -0.01 + 0.02i$ $a_{30} = -0.3 + 0.42i; a_{31} = -0.02 + 0.05i; a_{32} = -0.01 - 0.08i; a_{33} = 0.02 - 0.01i$	1	−39.1	−40.2	0.89
	$a_{10} = 1.4513 + 0.132i; a_{11} = -0.123 - 0.023i; a_{12} = 0.012 - 0.0043i$ $a_{30} = -0.132 - 0.430i; a_{31} = 0.322 + 0.243i; a_{32} = -0.0123 - 0.12i$ $a_{50} = -0.755 - 0.654i; a_{51} = -0.213 - 0.411i; a_{52} = 0.233 + 0.233i$	0.45	−41.2	−42.5	0.78
CGMP	$a_{10} = 0.9800 - 0.300i; a_{11} = 0.06 + 0.03i; a_{12} = 0.02 + 0.08i; a_{13} = -0.01 + 0.02i$ $a_{30} = -0.3 + 0.42i; a_{31} = -0.02 + 0.05i; a_{32} = -0.01 - 0.08i; a_{33} = 0.02 - 0.01i$	1	−56.1	−50.6	0.71
	$a_{10} = 1.4513 + 0.132i; a_{11} = -0.123 - 0.023i; a_{12} = 0.012 - 0.0043i$ $a_{30} = -0.132 - 0.430i; a_{31} = 0.322 + 0.243i; a_{32} = -0.0123 - 0.12i$ $a_{50} = -0.755 - 0.654i; a_{51} = -0.213 - 0.411i; a_{52} = 0.233 + 0.233i$	0.45	−57.1	−53.3	0.71

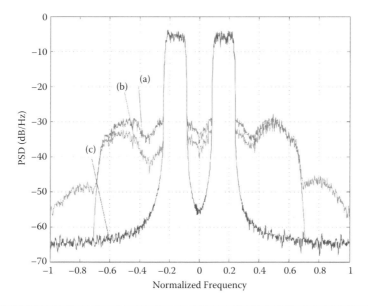

Figure 6.15 Power spectral density for two different power amplifiers with a two-carrier WCDMA signal applied: (a) power amplifier with $K = 2$ and $Q = 3$ and MEMR $= 1$; (b) power amplifier with $K = 3$ and $Q = 2$ and MEMR $= 0.45$; and (c) an input signal.

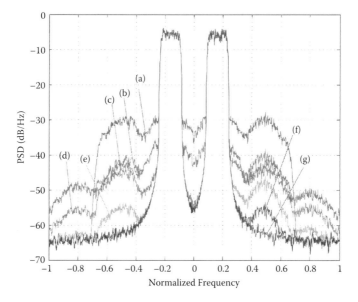

Figure 6.16 Comparison of the power spectral density between the memory polynomial predistorter and CGMP for a two-carrier WCDMA signal for a power amplifier with MEMR $= 1$: (a) output without predistortion; (b) output with memory polynomial predistortion ($Q = 3$, $K = 2$); (c) output with memory polynomial predistortion ($Q = 20$, $K = 2$); (d) output with CGMP (iteration $= 5$); (e) output with CGMP (iteration $= 7$); (f) output with CGMP (iteration $= 9$); and (g) input data.

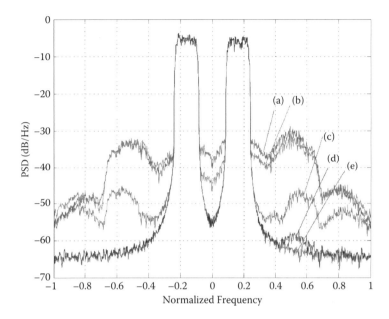

Figure 6.17 Comparison of the power spectral density between the memory polynomial predistorter and gain predistortion for a two-carrier WCDMA signal for a power amplifier with MEMR = 0.45: (a) output without predistortion; (b) output with memory polynomial predistortion ($Q = 0$, $K = 3$); (c) output with memory polynomial predistortion ($Q = 2$, $K = 3$); (d) output with CGMP (iteration = 5); and (e) input data.

Figure 6.17 shows the power spectral density when the power amplifier is with a MEMR = 0.45. According to Table 6.3, the ACLR for the memory polynomial technique is on average −41.8 dB and the EVM is 0.78%, and when applying the CGMP technique these values are −55 dB and 0.71%. In this case, the improvement of around 10 dB in ACLR is achieved. According to Figure 6.17, the memory polynomial technique could not suppress the out-of-band distortion even when the memory length increases to 20, but the CGMP technique completely reduces that effect. It should be noted that the number of iterations in the CGMP technique should be more when the amount of memory is more. The value of α in the adaptation equation should be between 0 and 1, and depending on how much backoff is applied, this value should be adjusted. With a higher value of α the convergence is faster, but it has the drawback of less efficiency, and the lower the value the convergence is slower but it has the advantage of reaching a higher output power. There is always a trade-off in choosing α.

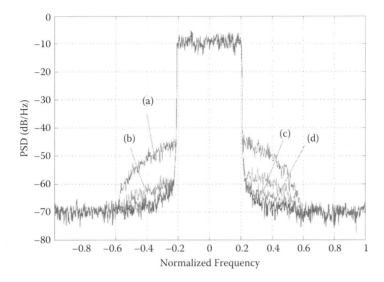

Figure 6.18 Comparison of the power spectral density between the memory polynomial predistorter and gain predistortion for a power amplifier with MEMR = 0.45 and Mobile WiMAX signal: (a) output without predistortion; (b) output with memory polynomial predistortion; (c) output with CGMP (iteration = 5); and (d) input data.

In Figures 6.18 and 6.19, the mobile WiMAX signal that is compatible with IEEE 802.16e with 10 MHz bandwidth is applied for simulations. The power amplifier is with a MEMR = 0.45 and the coefficients are shown in Table 6.3. According to Figure 6.18, the amount of reduction in ACLR is on average −49.3dB when applying the memory polynomial method and with the CGMP technique it is reduced to −57.4 dB, which is around an 8 dB improvement in ACLR as compared to the memory polynomial method. It is also shown that when the power amplifier is with MEMR = 1 it has more distortion than the one that has a MEMR = 0.45. This difference can be seen in Figures 6.18 and 6.19.

Figure 6.20 shows the EVM of the PA with predistortion for two power amplifiers for different iterations. The line that is marked with circles is the EVM of the PA with a MEMR = 0.45 and the line that is marked with x's is for the PA with a MEMR = 1.

This result shows that initially the power amplifier with more memory effects has an EVM that is more than the power amplifier with less memory effects after applying the CGMP and after two iterations the EVM drops to less than 1% and after more iterations the EVM can reach 0.71%. This is equal for both PAs.

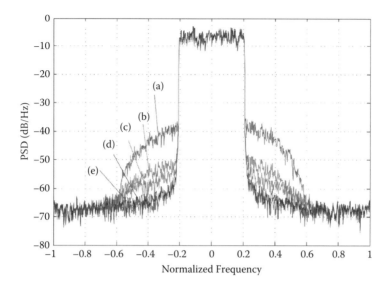

Figure 6.19 Comparison of the power spectral density between the memory polynomial predistorter and gain predistortion for a power amplifier with MEMR = 1 and a mobile WiMAX signal: (a) output without predistortion; (b) output with memory polynomial predistortion ($Q = 0$, $K = 3$); (c) output with memory polynomial predistortion ($Q = 2$, $K = 3$); (d) output with CGMP (iteration = 5); and (e) input data.

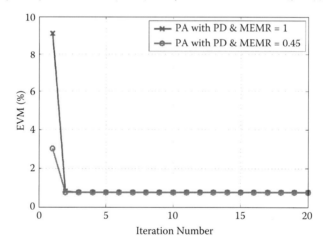

Figure 6.20 Error vector magnitude of the power amplifiers with different memory effects for the CGMP technique.

In Figure 6.21, the EVM is shown for the PAs but with applying the memory polynomial technique. This result is based on the memory length, which indicates the amount of memory in this technique. After increasing the memory length to 10, the final EVM is reached at 0.77%, which is close to the CGMP.

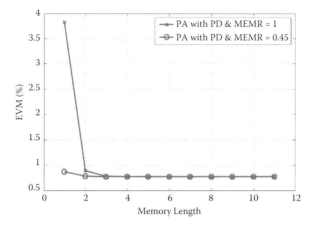

Figure 6.21   Error vector magnitude of the power amplifiers with different memory effects for the memory polynomial method.

6.4 Summary

In this chapter, a new predistortion technique is introduced that overcomes the problem of memory effects of the power amplifier, especially the dynamic effects or electrical memory effects. This new technique was first proven with equations that were based on the Volterra series and complex gain predistortion, and a simulation demonstrated the validity of this technique. Then two algorithms were discussed for implementing this technique and the advantages and disadvantages of each are studied. Two power amplifiers are applied for simulation and each one has different memory effects. The MEMR is used to elaborate the amount of memory in the power amplifier that is designed. The power amplifier that is designed here and used in the simulations is MRF1806 from Motorola. After modeling this power amplifier with memory, the MEMR is equal to 0.45. The other power amplifier is the one that was designed by Ku and Kenney (2003) and has a MEMR = 1. The simulations are done for both PAs to show the effectiveness of the CGMP technique. Two input signals are used for doing the simulation. The improvement of 25 dB to 30 dB is achieved when applying the CGMP technique. This is around a 13 dB improvement when applying the memory polynomial technique. The PSD is plotted for WiMAX and two-carrier CDMA signals. Also, the reduction in EVM is acceptable.

In the next chapter, the implementation for digital predistortion is presented in detail using Xilinx blocksets in MATLAB.

References

Black, H. S. 1997. Inventing the negative feedback amplifier. *IEEE Spectrum* 14:55–60.

Bondar, D., and Budimir, D. 2009. Digital baseband predistortion of wideband power amplifiers with improved memory effects. *RWS 2009 IEEE Radio and Wireless Symposium, Proceedings* 4957334:284–287.

Bosch, W., and Gatti, G. 1989. Measurement and simulation of memory effects in predistortion linearizers. *IEEE Transactions on Microwave Theory and Techniques* 37(12).

Boumaiza, S., and Ghannouchi, F. 2003. Thermal memory effects modeling and compensation in RF power amplifiers and predistortion linearizers. *IEEE Transactions on Microwave Theory and Techniques* 51(12).

Carvalho, N. B., and de Pedro, J. C. 2002. *IEEE Transactions on Microwave Theory and Techniques* 50(9).

Cavers, J. 1990. Amplifier linearization using a digital predistorter with fast adaptation and low memory requirements. *IEEE Transactions on Vehicular Technology* 39(4).

Cavers, J. K. 1997. The effect of quadrature modulator and demodulator errors on adaptive digital predistorters for amplifier linearization. *IEEE Transactions on Vehicular Technology* 46(2):456–466.

Clark, C. J., Chrisikos, G., Muha, M. S., Moulthrop, A. A., and Silva, C. P. 1998. Time-domain envelope measurement technique with application to wideband power amplifier modeling. *IEEE Transactions on Microwave Theory Technology* 46:2531–2540.

Cripps, S. C. 1999. *RF Power Amplifiers for Wireless Communications*. Norwood, MA: Artech House.

Ding, L., Zhou, G. T., Morgan, D. R., Ma, Z., Kenney, J. S., Kim, J., and Giardina, C. R. 2004. A robust predistorter constructed using memory polynomials. *IEEE Transactions on Communications* 52(1):159–165.

Eun, C., and Powers, E. J. 1997. A new Volterra predistorter based on the indirect learning architecture. *IEEE Transactions on Signal Processing* 45:223–227.

Franco, M. Minimizing power amplifiers memory effects. 2004 (June). IEEE MTT International Microwave Symposium Workshop on Distortion Correction of High Power Amplifiers Using Digital Signal Processing, Dallas, TX.

Gilabert, P. L., Bertran, E., Montoro, G., and Berenguer, J. 2004. Study on the robustness of a 22 MHz bandwidth feedforward amplifier at the 2.4 GHz ISM-band. *Proceedings of IEEE Personal, Indoor and Mobile Radio Communications* (PIMRC '04) 1:186–190.

Hetzel, S. A., Bateman, A., and Mcgeehan, J. P. 1991. LINC transmitter. *Electronic Letters* 27(10):133–137.

Jeckeln, E., Shih, H., Martony, E., and Eron, M. 2003. Method for modeling amplitude and bandwidth dependent distortion in nonlinear RF devices. *IEEE MTT International Microwave Symposium Digest.*

Kenington, P. B. 2000. *High-Linearity RF Amplifier Design*. Norwood, MA: Artech House.

Kim, J., and Konstantinou, K. 2001. Digital predistortion of wideband signals based on power amplifier model with memory. *Electronic Letters* 37:1417–1418.

Ku, H., and Kenney, J. 2003. Behavioral modeling of nonlinear RF power amplifiers considering memory effects. *IEEE Transactions on Microwave Theory and Techniques* 51(12).

Ku, H., and McKinley, M. D. 2002. Quantifying memory effects in RF power amplifiers. *IEEE Transactions on Microwave Theory and Techniques* 50(12):2843–2849.

Liu, Y. J., Chen, W., Zhou, J., Zhou, B. H., and Ghannouchi, F. 2013. Digital predistortion for concurrent dual-band transmitters using 2-D modified memory polynomialism. *IEEE Transactions on Microwave Theory and Techniques* 61(1):281–290.

Ma, Z., Zierdt, M., and Pastalan, J. 2001. Characterization of power amplifier memory effect for digital baseband predistortion. Unpublished work, January.

Maas, S. A. 1997. *Nonlinear Microwave Circuits*. Piscataway, NJ: IEEE Press.

Mato, J. L., Pereira, M., Rodríguez-Andina, J. J., Fariña, J., Soto, E., and Pérez, R. 2009. FPGA-based implementation of segmented predistorters for RF power amplifiers. *IEEE International Symposium on Industrial Electronics* 4677314:1948–1952.

Morgan, D. R., Ma, Z., Kim, J., Zierdt, M. G., and Pastalan, J. 2006. A generalized memory polynomial model for digital predistortion of RF power amplifiers. *IEEE Transactions on Signal Process* 54(10):3852–3860.

Patel, J. 2004 (March). Adaptive Digital Predistortion Linearizer for Power Amplifiers in Military UHF Satellite. Department of Electrical Engineering, University of South Florida.

Pothecary, N. 1999. *Feedforward Linear Power Amplifiers*. Norwood, MA: Artech House.

Schetzen, M. 1980. *The Volterra and Wiener Theories of Nonlinear Systems*. New York: Wiley.

Sundstrom, L. 1996. The effects of quantization in digital signal component separator for LINC transmitter. *IEEE Transactions on Vehicular Technology* 45(2):346–352.

Woo, Y. Y., Kim, J., Hong, S., Kim, I., Moon, J., Yi, J., and Kim, B. 2007. Adaptive digital feedback predistortion technique for linearizing power amplifiers. *IEEE Transactions on Microwave Theory Technology* 55(5):932–940.

Wright, A., and Nesper, O. 2002. Multi-carrier WCDMA base station design considerations amplifier linearization and crest factor control. Technology White Paper, PMC-Sierra, Santa Clara, CA.

Vuolevi, J. 2003. *Distortion in RF Power Amplifiers*. Artech House, Inc.

Zhe, J., Zhihuan, S., and Jiming, H. 2008. Volterra series based predistortion for broadband RF power amplifiers with memory effects. *Journal of Systems Engineering and Electronics* 19(4):666-667.

7

DIGITAL PREDISTORTION IMPLEMENTATION

7.1 Introduction

In this chapter, the implementation procedure of digital predistortion as a promising technique for simulation in MATLAB and using Xilinx blocksets toward compiling a field programmable gate array (FPGA) board is introduced. All these procedures are done in MATLAB and also with Integrated Software Environment (ISE) software from Xilinx. The Xilinx blockset tools are discussed and show the FPGA board from Xilinx that is used for the implementation. The implementation details are shown for each part of the predistortion filter that includes a complex multiplier, lookup tables (LUTs), and a complex divider. Each of these parts is designed and explained separately and the most important one for implementing the complex gain memory predistortion (CGMP) technique—the complex divider—is discussed in detail. A new algorithm for implementing the complex divider is proposed that is based on the Newton-Raphson method. The complex divider is compared with the other divider algorithms in terms of hardware resources and divider error. After that the CGMP block is created and using a Xilinx System Generator the VHDL (VHSIC Hardware Description Language) code is generated and ready to compile in FPGA. To make the testing easier, the other block, which is the JTAG Co-Simulation, is also created to help the compilation and real-time testing that can be done in MATLAB. Finally, the results of this implementation are presented.

7.2 Simulation with Xilinx Blocksets

In recent years, FPGAs have become a fundamental component in implementing high-performance digital signal processing (DSP) systems, especially in the areas of digital communications, networking, and video. The logic fabric of FPGAs consists not only of lookup tables, registers, multiplexers, and memory, but also dedicated circuitry for fast adders, multipliers, and I/O processing. The memory of a FPGA is more than that of a microprocessor or DSP processor, but DSP runs at clock rates 2 to 10 times that of the FPGA. With the capability for implementing high arithmetic architectures, this makes the FPGA ideally suited for creating high-performance custom data path processors for tasks such as digital filtering, fast Fourier transforms, and forward error correction. A third-generation (3G) wireless base station typically contains FPGAs and ASICs (application-specific integrated circuits) in addition to microprocessors and DSPs.

System Generator is a software tool for both modeling and designing FPGA-based DSP systems in Simulink. This tool gives a high-level implementation of a DSP system. With this tool, first the design is simulated and with the features of this tool the whole design can be compiled in VHDL or Verilog codes. It is even possible to make a project file and load it in the ISE software for more analysis or to modify the code (Xilinx Guides, 2008).

7.2.1 System Generator

Simulink provides a powerful high-level modeling environment for DSP systems, and then it is widely used for algorithm development and verification. The implementation is easy in that the system model and hardware implementation are bit-identical and cycle-identical at sample times defined in Simulink. The implementation is made efficient through the instantiation of intellectual property (IP) blocks that provide a range of functionality from arithmetic operations to complex DSP functions. These IP blocks are carefully designed to run at high speed and to be area efficient. User-defined IP blocks can be incorporated into a System Generator model as black boxes, which will be embedded by the tool into the HDL implementation of the design (Xilinx Guides, 2008).

7.3 Xilinx Embedded Development Kit

The Xilinx Embedded Development Kit (EDK) includes Embedded System Tools (EST), documentation, and hardware IPs for the Xilinx-embedded processors and peripherals (Xilinx Guides, 2008).

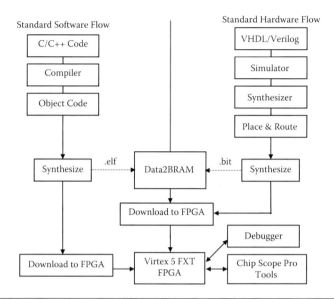

Figure 7.1 Flowchart for the system design flow using the Xilinx EDK. (From Lin, C. H. et al. 2006. *IEEE Transactions on Microwave Theory and Techniques* 54:2118–2127. With permission.)

Every embedded system design using the Xilinx EDK is divided into two parts: hardware design and software design.

Once the hardware design is implemented and the software programs are compiled, they can be combined into a bit stream and then downloaded to the target system. Figure 7.1 is a detailed block diagram that shows the system design flow using Xilinx EDK. If the software program uses on-chip block RAMs, the object code (.elf) can be combined with the hardware bit stream (.bit) to form a new downloadable bit stream using the DATA2BRAM utility.

DATA2BRAM then takes the .bit file as an input and adds new block random access memory (RAM) contents. This will eliminate the need to reimplement the entire system after modifications have been made to the software. To debug a software program, the software debugger also can be used that comes with the Xilinx EDK. Hardware debugging requires Xilinx ChipScope, if no other device such as an oscilloscope or a logic analyzer is available.

7.4 Field Programmable Gate Array

Field programmable gate arrays (FPGAs) and microprocessors accomplish different tasks. FPGAs are configurable and reprogrammable

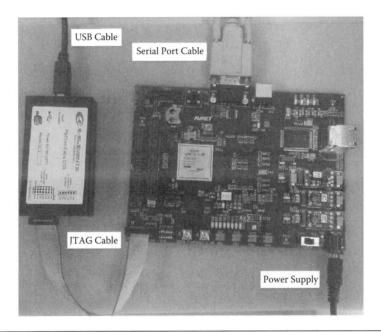

Figure 7.2   The Virtex-5 FXT Evaluation board.

digital logic devices, and programming code is written in Hardware Description Languages (HDL). Microprocessors execute predefined commands and do not have much flexibility. Engineers usually write programs for microprocessors in a language such as C.

Here, the Virtex-5 FXT Evaluation Kit is used, which has the Xilinx Virtex-5 XC5VFX30T-FF665 FPGA chip. This board also has 64 MB DDR2 SDRAM memory and 16 MB flash memory with a variety of I/O gates. A picture of this board is shown in Figure 7.2. It can be seen that the JTAG cable is connected to the JTAG connector of the board for programming the FPGA and the RS232 serial port for programming the flash memory and PowerPC processor with the use of EDK software.

7.4.1 Description

The Virtex-5 FXT Evaluation Kit provides a complete hardware environment to accelerate their time to market. The installed Virtex-5 FX30T device offers a prototyping environment to effectively demonstrate the enhanced benefits of Xilinx FPGA solutions. Reference designs are included with the kit to exercise standard

Table 7.1   XC5VFX30T Features

DEVICE	NUMBER OF SLICES	BLOCK RAM (KB)	DSP48E SLICES
XC5VFX30T	5120	2448	64

peripherals on the development board for a quick start to device familiarization.

7.4.2 Functional Description

The Virtex-5 FX30T FPGA features four digital clock managers (DCMs), two phase lock loops (PLLs), and 1.25 Gbps LVDS I/O. Table 7.1 shows some other main features of the FF676 package.

7.5 Complex Gain Memory Predistortion Implementation

Here, the target is to implement the complex gain memory predistortion (CGMP) technique that is proposed in Chapter 6:

$$F_{i+1}\left(|x(n)|^2\right) = F_i\left[|x(n)|^2\right] - a\frac{F_i\left[|x(n)|^2\right]}{v(n)G_0\left[|v(n)|^2\right]}V_{error}(n) \tag{7.1}$$

It is discussed in two cases and with strong reasons, the second case is chosen, which is

$$F_{i+1}\left(|x(n)|^2\right) = F_i\left[|x(n)|^2\right] - a\frac{1}{Gx(n)}V_{error}(n) \tag{7.2}$$

This implementation should be a feasibility study. Therefore, the number of multipliers and block RAMs should be in the range of hardware resources, as it is limited for the FPGA that is used here, especially for finding the inverse of the $Gx(n)$ for Equation (7.2). The most difficult part of the algorithm is the implementation of the division. There are no hard macros for dividers on the target chip. So the divider should be designed with other Xilinx blocks that are available in FPGA a like a multiplier, adder, and subtracter. A complex divider should be designed with a combination of these blocks.

The other important issue during the implementation of the CGMP technique is the clock allocation for this block. This means that the CGMP runs with a single clock or multiple clocks, which in terms of computation rate for multiple clocks, is faster. Here, the single clock system is applied for all the parts, but for the best performance and high efficiency of the system, separate clocks should be applied to minimize a delay in the system. CGMP includes different parts as was discussed in detail in Chapter 6. The main parts are the complex multiplier block; LUT for storing predistortion data, which here block; RAMs are used; and the divider. These parts are explained in detail next.

7.5.1 Complex Multiplier

The complex multiplier block actually multiplies the complex values of input data with the complex values from the LUT. Figure 7.3 shows the Xilinx blocks for this multiplier.

According to Figure 7.3, the multiplication is designed based on the following:

$$(Xi + j \times Xq) \times (Yi + j \times Yq) = (Xi \times Yi - Xq \times Yq)$$
$$+ j \times (Xi \times Yq + Xq \times Yi) \quad (7.3)$$

Then the output result is represented as

$$Pi = (Xi \times Xq - Yi \times Yq)$$
$$Pq = (Xi \times Yq + Xq \times Yi) \quad (7.4)$$

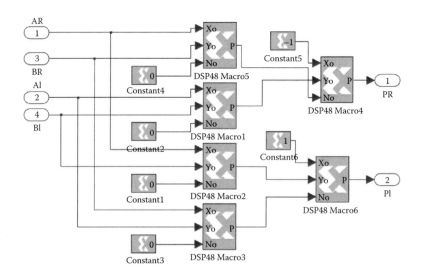

Figure 7.3 Complex multiplier implementation with Xilinx blocks.

According to the multiplier implementation, the hardware resource is 4 multipliers out of 64 that are available in Virtex-5, and one addition and one subtraction.

7.5.2 Lookup Table (LUT)

In CGMP, two LUTs store the complex I and Q values. For implementing LUT a single port RAM is used. The Xilinx single port RAM block implements RAM, as shown in Figure 7.4.

The block has one output port and three input ports for address, input data, and write enable (WE). Values in a single port RAM are stored by word, and all words have the same arithmetic type, width, and binary point position.

The block has two types of implementations, using either block or distributed memory. Each data word is related with exactly one address that can be any unsigned integer from 0 to d–1, where d denotes the RAM depth (number of words in the RAM). The initial RAM content can be specified through the block parameters.

The write enable port must be a 1-bit unsigned integer. When the WE port is 1, the value on the data input is written to the location indicated by the address line. The output during a write operation depends on the choice of memory. For distributed memory, the output port always has the value at the location specified by the address line. For block memory, the behavior of the output port depends on the write mode selected. When the WE is 0, the output port has the value at the location specified by the address line.

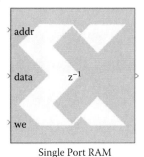

Single Port RAM

Figure 7.4 Single port RAM block. (From Xilinx Guides. 2008. System Generator for DSP, Getting Started Guide, Release 10.1. With permission.)

Parameters specific to this block are

Depth—Specifies the number of words stored; must be a positive integer.

Initial value vector—Specifies the initial value.

Zero initial output—When checked, the data out ports have a value of zero at clock 0; otherwise, they have a value of NaN ("not a number").

Write mode—Specifies the memory behavior to be Read Before Write, Read After Write, or No Read On Write. There are device-specific restrictions on the applicability of these modes.

Use distributed memory (instead of block RAM)—When checked, the block is implemented with distributed RAM; otherwise, it is implemented with block RAM.

7.6 Complex Divider Implementation

There is no divider block inside the FPGA. This block should be designed by the other existing blocks like a multiplier, adder, and subtracter, and because it can be implemented in several different ways; there is no specific algorithm that is clearly the one to choose. It all depends on the requirements one have on its properties, such as accuracy, size, and speed. The implementation is expected to be a part of a baseband processor and should be able to handle the high speed requirements while keeping the size down.

There are two ways to perform a division with logical circuitry. The way that is used by the Xilinx CoreLib is to calculate each bit exactly. It is based on the CORDIC algorithm. The other method is based on the Newton-Raphson method. The following sections describe the complex divider algorithm and how to implement it. After that, the Newton-Raphson method and how it is used for approximating a division is explained. Simulink is used to optimize the algorithm and to analyze the error. Finally, the Xilinx block of the complex divider is shown and the results are also presented.

The details of the implementation of the complex divider were discussed in Chapter 5.

Here, the complete Xilinx simulation of the CGMP is shown. In Figure 7.5, the final predistortion filter block is shown. According to Figure 7.5, this block has six inputs and four outputs.

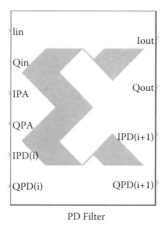

Figure 7.5 CGMP block.

The Iin and Qin are the main inputs from the source. The other inputs of this block are IPA and QPA, which are the feedbacks that are coming from the power amplifier. IPD(i) and QPD(i) are the values in the LUT. It is assumed that these values are initially 1 and 0, accordingly. The Iout and Qout are the main outputs that are sent to the power amplifier and the IPD(i+1) and QPD(i+1) are connected to the two IPD(i) and QPD(i), respectively. With the Xilinx System Generator block the VHDL code can be created from the final predistortion block.

Now the JTAG Co-Simulation block is generated in order to be compiled into the FPGA. Figure 7.6 is the result of this. It will take 10 minutes in the Pentium 4 PC with 2 gigabyte RAM to generate it. In Table 7.2, the hardware resources that are used for implementing this block are shown.

The clock period is 25 MHz, which is the clock period for JTAG Co-Simulation. The other clock period for PCI could be 133 MHz and for the Ethernet, it is a maximum of 1 GHz. In the Virtex-5 FXT board there is no Ethernet Co-Simulation block and that is why JTAG is used for implementing in the FPGA.

The complete simulation of the CGMP with the power amplifier and the input signal is shown in Figure 7.9. The input signal here is created in Microwave Office 2006 and then imported in MATLAB and is shown as I and Q in this figure. After running this simulation, the CGMP is generating LUT values that are stored in block RAMs.

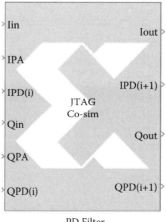

Iin	Iout
IPA	
IPD(i)	IPD(i+1)
	JTAG Co-sim
Qin	Qout
QPA	
QPD(i)	QPD(i+1)

PD Filter

Figure 7.6 JTAG Co-Simulation block of the CGMP.

Table 7.2 Hardware Resources of the Predistortion Filter

XC5VFX30T-1FF665	RESOURCES USED	PERFORMANCE
Slices	464	9%
DSP48 slices	35	54%
Number of fully used LUT–FF pairs	463	88%
Number of block RAM/FIFO	12	17%

The input signal passes through the PA and then the feedback from the PA and samples from the input signal are sent to the predistortion block where these samples are stored in RAMs to update the LUT. In FPGA, these samples are in first input first output (FIFO). After some iteration, the output samples from the predistortion block are generating data that after passing through the PA produces the spectrum that is without distortions. In CGMP, there is also an α parameter, which is the gain factor that can control the adaptation process. Figure 7.10 shows the Xilinx simulation of the inside of the predistortion block. As shown in Figure 7.10, there is a divider block, complex multiplier, and RAM blocks, which were discussed earlier in this chapter.

7.7 Results of FPGA Implementation

Figure 7.7 shows the result of the comparison between implementing CGMP in FPGA and simulation results that were discussed before. In Figure 7.7, the spectrum of the output and input of the

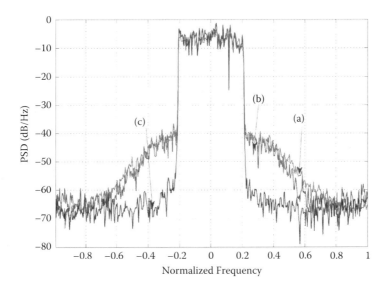

Figure 7.7 Comparison of PSD between simulation and implementation results: (a) implementation result after applying CGMP with one iteration, (b) simulation result after applying CGMP with one iteration, and (c) input signal.

PA before applying the CGMP technique is shown. This result captures 1024 samples; the LUT size is 10 bits. The result of implementation, point (a) in Figure 7.7, is almost equal to the simulation results shown in point (b). To obtain these results, the JTAG Co-Simulation block is created from the CGMP that was designed previously. Then a block series with a power amplifier are applied for real-time implementation.

The JTAG block compiles the CGMP into FPGA through the JTAG cable. Then, samples from MATLAB, which are stored in the workspace, are transmitted to the FPGA through RAM blocks that represent as FIFO. Then, the LUT values, which are initially one multiple with the received values, and these samples are sent from the FPGA to the PC and then pass to the PA. From the PA, the data is fed back to the JTAG block. These values update the LUTs that are I and Q values. After some iteration, the nonlinearity of the power amplifier is reduced as shown in Figures 7.7 and 7.8.

In Figure 7.8 the result after applying the CGMP technique is shown and is compared to the simulation results. It can be seen that the results are equal but the simulation result is slightly better than the implementation result. It should be noted that in FPGA

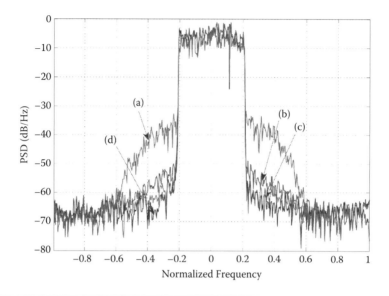

Figure 7.8 Comparison of PSD between simulation and implementation results: (a) without predistortion, (b) implementation result after applying CGMP with five iterations, (c) simulation result after applying CGMP with five iterations, (d) input signal.

implementation because of the constraint in FPGA resolution, the result of the divider implementation as shown in Figure 7.8 is slightly different than the simulation result. Point (a) is when the power amplifier is without predistortion; point (b) is the implementation result after applying the CGMP technique and after five iterations. The reduction between 15 dB to 20 dB is achieved from the implementation results, which is less than the simulation result, which has the reduction between 20 dB to 25 dB in adjacent channel leakage ratio (ACLR).

In Figure 7.9, the complete simulation of the predistortion block with PA is shown. The predistortion block contains the CGMP technique and series with the PA that shows nonlinearity. The input signals are shown with I and Q blocks. As discussed earlier in this chapter, the predistortion is designed based on the complex signal and then the input should also be in Cartesian form. The block for generating the VHDL code is also present in the simulation. Figure 7.10 shows the inside of the predistortion block and contains a complex multiplier, complex divider, and block RAMs. Also, some delay blocks that are using for compensating any delay between the path of the transmission and between blocks are available.

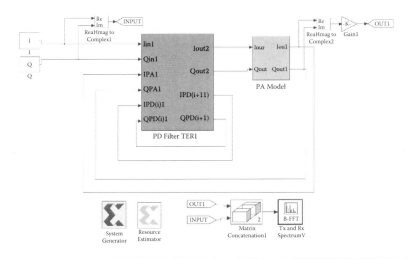

Figure 7.9 CGMP simulation with a power amplifier.

7.8 Digital Signal Processing Implementation of Digital Predistortion

In Figure 7.11, it is observed that the digital predistortion (DP) block of the design requires both the input and output of the system in terms of amplitude (S) and phase (R). After processing the input and output required, the processed or predistorted input will be fed into the power amplifier.

The output from the power amplifier is again fed into the DP block in order to prepare the block for the next predistortion process. The number of iterations required to eliminate distortion and linearizing the power amplifier may vary for different signal types.

Figure 7.12 shows the general implementation flow, starting from the top level, which is a DP block design, to the next layer, which is a software implementation in MATLAB and C++, and last the lowest layer, which is the hardware implementation that uses the Embedded Development Kit (EDK) developed by Xilinx.

7.9 DP Block Design

The DP block contains multiple blocks. The adaptation algorithm design will be discussed first. Next the design discussion will be brought to the adaptation block, which is the key block of the design. Blocks used for the purpose of testing and experimenting such as the complex multiplier, Saleh model amplifier, and the

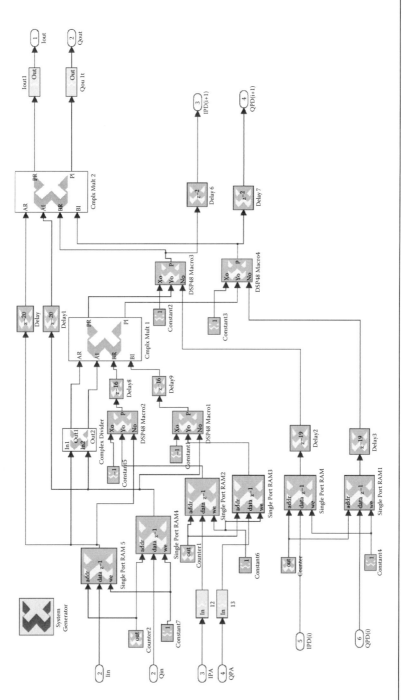

Figure 7.10 CGMP implementation with Xilinx blocksets.

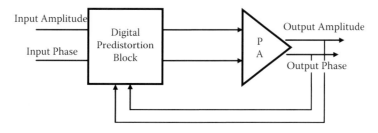

Figure 7.11   General project design for digital predistortion.

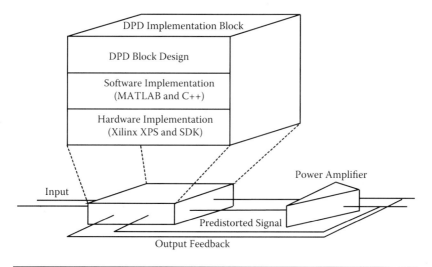

Figure 7.12   Implementation design diagram for DP PA linearization.

in-phase quadrature scale rotate (IQSR) block will also be designed. Figure 7.13 summarizes the available components for DP.

7.9.1 Linear Convergence Adaptation Algorithm

The linear convergence adaptation algorithm used in the project design is as follows:

$$S_{n+1} = S_n - (\alpha \times V_{errA}) \tag{7.5}$$

$$R_{n+1} = R_n - (\beta \times V_{errP}) \tag{7.6}$$

V_{errA} is the amplitude difference between the input signal and the output signal.

$$V_{errA} = S_{output} - S_{input} \tag{7.7}$$

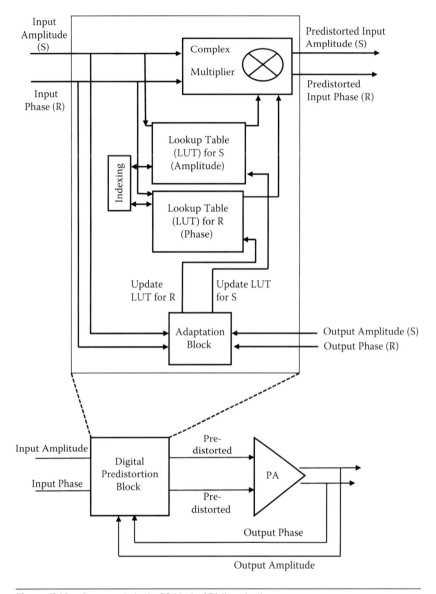

Figure 7.13 Components in the DP block of PA linearization.

V_{errP} is the phase difference between the input signal and the output signal in radians.

$$V_{errP} = R_{output} - R_{input} \qquad (7.8)$$

Alpha and *beta* is the parameter used to control the amplitude and phase of the output.

S is the amplitude and R is the phase of the signal in terms of radians.

The complex representation in polar form can be expressed as follows:

$$Z = I + Qi \tag{7.9}$$

$$Z = Scos(R) + Ssin(R)i \tag{7.10}$$

The S and R will remain unchanged if $VerrA$ or $VerrP$ is zero (no difference in amplitude and phase between input and output). The S and R values are then stored in the LUT to be multiplied with the input signal by the complex multiplier. The linear convergence adaptation algorithm is chosen due to its simplicity in performing operations unlike other methods, which require a high level of calculation thus further burdening the DSP (Varahram, 2009).

7.9.2 Adaptation Block

The flowchart in Figure 7.14 explains an updating process that happens in the adaptation block. First the S and R are initialized with 1s and 0s. After the input and output signal is acquired and fed into the block, the indexing block starts to initialize indexing the LUT for S and R using the input amplitude value as a reference.

The first updated LUT block has to be located outside the iterations loop, as there is still a need to index the LUT after the first LUT update. The update starts by acquiring the amplitude and phase differences between the input and output signals, and later that data is used to update the LUT of S and R by using the linear convergence adaptation algorithm. The update will occur at a fast rate, as the adaptation does not use complex calculations but only simple addition and multiplication operations.

After the first update is done, the indexing block reacts again by indexing the LUT of S and R with the index table that was generated previously during the initialization of the indexing block. The process then leads to the iteration loop.

The iteration loop continues until the maximum iteration. The block acquires new output data at each new iteration loop but does not go through the indexing process again.

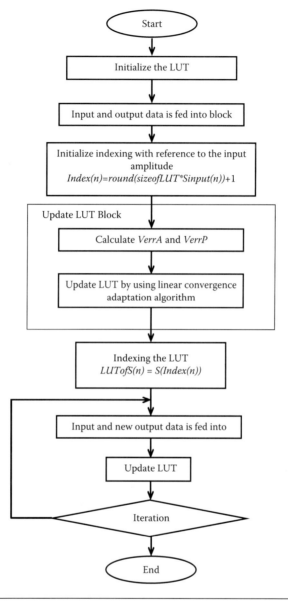

Figure 7.14 Flowchart for the adaptation block operation.

7.9.3 Complex Multiplier

Figure 7.15 shows the design for the complex multiplier block used to verify the operation of the adaptation block in the project. The complex multiplier has two multiplier components in it, one for the *S* (amplitude) and one for the *R* (phase) of the predistortion.

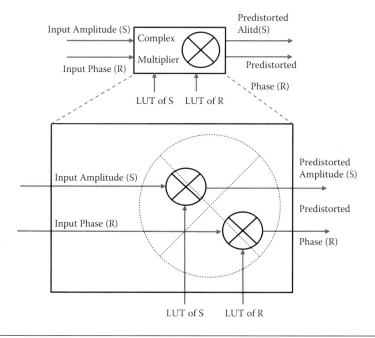

Figure 7.15   Complex multiplier design.

The S of the input signal is multiplied with the S found in the LUT and the same goes for R. Since the LUT is already indexed in the adaptation block, the multiplication is carried out without the need to manually select data from the LUT, as they were already arranged properly from indexing.

7.9.4 Saleh Model Amplifier

The Saleh model amplifier is used in this project to test the workability of the adaptation algorithm. The input to the amplifier is tuned to reach its nonlinear region in order to see how well the adaptation block design is able to suppress the distortion and also linearize the amplifier. The characteristics of the Saleh model amplifier are summarized in the following equations:

$$Output\ (amplitude) = \frac{alpha1 \times u}{1 + (beta1 \times u^2)} \tag{7.11}$$

$$(Output\ (phase) = Input\ (phase) + \frac{alpha2 \times u^2}{1 + (beta2 \times u^2)} \tag{7.12}$$

where alpha1 = 2.1587, alpha2 = 4.0033, beta1 = 1.1517, beta2 = 9.1040, and u = Input (amplitude). The gain of the Saleh model amplifier will be 2.1587 (Varahram, 2009).

7.9.5 The IQSR Block

The in-phase quadrature scale rotate (IQSR) block is used to convert complex signals into the data form that is needed by the other blocks in the DP power amplifier linearization (Figure 7.16).

$$Z => I + Qi \text{ or } Z => S/R$$

Table 7.3 shows the performance difference for a two-carrier CDMA (code division multiple access) signal of a pre-amplifier gain of 0.05. Table 7.4 shows the performance difference for a two-carrier CDMA signal of a pre-amplifier gain of 0.10. Table 7.5 shows the performance difference for a two-carrier CDMA signal of a

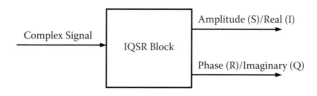

Figure 7.16 IQSR block design.

Table 7.3 ACPR Reduction for a Two-Carrier CDMA Signal of a Pre-Amp Gain of 0.05

ITERATIONS	ACPR, MATLAB ($ALPHA = 0.4$, $BETA = 0.4$)	ACPR, C++ ($ALPHA = 0.4$, $BETA = 0.4$)
5	10 dB	11 dB
10	12 dB	16 dB
20	18 dB	20 dB

Table 7.4 ACPR Reduction for a Two-Carrier CDMA Signal of a Pre-Amp Gain of 0.10

ITERATIONS	ACPR, MATLAB ($ALPHA = 0.4$, $BETA = 0.4$)	ACPR, C++ ($ALPHA = 0.4$, $BETA = 0.4$)
5	12 dB	7 dB
10	19 dB	18 dB
20	24 dB	22 dB

Table 7.5   ACPR Reduction for a Two-Carrier CDMA Signal of a Pre-Amp Gain of 0.20

ITERATIONS	ACPR, MATLAB ($ALPHA = 0.4$, $BETA = 0.4$)	ACPR, C++ ($ALPHA = 0.4$, $BETA = 0.4$)
5	12 dB	—
10	20 dB	—
20	35 dB	—

Table 7.6   ACPR Reduction for a Two-Carrier CDMA Signal of a Pre-Amp Gain of 0.40

ITERATIONS	ACPR, MATLAB ($ALPHA = 0.4$, $BETA = 0.4$)	ACPR, C++ ($ALPHA = 0.4$, $BETA = 0.4$)
5	—	—
10	20 dB	—
20	50 dB	—

pre-amplifier gain of 0.20. There is no value in the C++ column because there is no reduction in the adjacent channel power ratio (ACPR) for a pre-amp gain of 0.20 in the C++ system. This indicates that the C++ system has reached its limit. Table 7.6 shows the performance difference for a two-carrier CDMA signal of a pre-amplifier gain of 0.40.

The C++ column remains valueless, as it has reached its limit in the previous 0.2 pre-amp gain. The MATLAB simulation for five iterations shows no value because the number of iterations is insufficient to achieve significant reduction. The table for a pre-amp gain of 0.8 is not shown here because both systems have reached their performance limit at 0.8.

The following tables are the results of using the Worldwide Interoperability for Microwave Access (WiMAX) signal by varying the pre-amp gain and the number of iterations. Table 7.7 shows the performance difference for WiMAX signal of a pre-amplifier gain of 0.07. Table 7.8 shows the performance difference for the WiMAX signal of a pre-amplifier gain of 0.15. Table 7.9 shows the performance difference for the WiMAX signal of a pre-amplifier gain of 0.30.

There is no value in the C++ column in Table 7.9 because there is no reduction in ACPR for a pre-amp gain of 0.30 in the C++ system. This indicates that the C++ system has reached its limit.

Table 7.10 shows the performance difference for a two-carrier CDMA signal of a pre-amplifier gain of 0.60. The C++ column remains valueless as it has reached its limit in the previous 0.3 pre-amp gain. The MATLAB simulation for 20 iterations shows no value

Table 7.7 ACPR Reduction for a WiMAX Signal with a Pre-Amp Gain of 0.07

ITERATIONS	ACPR, MATLAB ($ALPHA = 0.4$, $BETA = 0.4$)	ACPR, C++ ($ALPHA = 0.4$, $BETA = 0.4$)
5	2 dB	2 dB
10	2 dB	2 dB
20	2 dB	2 dB

Table 7.8 ACPR Reduction for a WiMAX Signal with a Pre-Amp Gain of 0.15

ITERATIONS	ACPR, MATLAB ($ALPHA = 0.4$, $BETA = 0.4$)	ACPR, C++ ($ALPHA = 0.4$, $BETA = 0.4$)
5	10 dB	8 dB
10	15 dB	15 dB
20	15 dB	16 dB

Table 7.9 ACPR Reduction for a WiMAX Signal with a Pre-Amp Gain of 0.30

ITERATIONS	ACPR, MATLAB ($ALPHA = 0.4$; $BETA = 0.4$)	ACPR, C++ ($ALPHA = 0.4$, $BETA = 0.4$)
5	10 dB	—
10	17 dB	—
20	20 dB	—

Table 7.10 ACPR Reduction for a WiMAX Signal with a Pre-Amp Gain of 0.60

ITERATIONS	ACPR, MATLAB ($ALPHA = 0.4$, $BETA = 0.4$)	ACPR, C++ ($ALPHA = 0.4$. $BETA = 0.4$)
5	10 dB	—
10	10 dB	—
20	—	—

because the MATLAB system has reached its performance limit and no reduction in ACPR is observed.

The preceding tables have shown that the ACPR reduction could be improved by increasing the number of iterations for each system provided that the pre-amp gain is within the allowable range.

It is also discovered that the MATLAB simulation performs better than the C++ implementation, due to the fact that MATLAB has higher and better resolution compared to C++. Therefore, the calculations are more accurate, causing the predistortion to be more effective. The MATLAB system is able to operate relatively nearer to the nonlinear region compared to the C++ system.

Table 7.11 Maximum Pre-Amp Gain Allowable for MATLAB and C++ Systems with Respect to Signal Types

SIGNAL TYPE/SYSTEM TYPE	MATLAB SIMULATION	C++ IMPLEMENTATION
Two-carrier CDMA	0.6	0.11
WiMAX	0.55	0.15

Table 7.11 shows the maximum allowable pre-amp gain for the MATLAB simulation and C++ implementation at their respective signal type input. It is observed from the table that the maximum allowable pre-amp gain for the two signal types are different. This is because WiMAX, which belongs to the 4G network, has a higher bandwidth; therefore a higher peak-to-average power ratio (PAPR) if compared to the two-carrier CDMA, which belongs to the 3G network. A higher PAPR means that the WiMAX signal needs to be backed-off further compared to the two-carrier CDMA signal, which results in a smaller allowable pre-amp gain. The C++ has only about 20% of the MATLAB simulation performance due to smaller operating resolutions compared to MATLAB.

7.10 Summary

The experiments conducted in this project use the example signals from 3G and 4G, which is a two-carrier CDMA signal and WiMAX signal. The signals are extracted from the ADS available at the Engineering Tower, UPM. The experiment site is set up using the Xilinx XtremeDSP Development Kit-4 in which the DSP Microblaze Processor is built with it.

The MATLAB simulation is performed by feeding the system with the aforementioned signals and its output is plotted in a two-performance graph, the power spectral density (PSD) graph, AM-AM graph, and the AM-PM graph. The C++ coding, which could be programmed directly onto the Microblaze Processor, is also fed with the respective signals to test the operability range of the system.

In order to control the experiment, the parameters involved are the pre-amp gain, the number of iterations, and the constants *alpha* and *beta*. The range of allowable pre-amp gain for each system is different and summarized in tables by conducting experiments. The higher the number of iterations, the better the performance of the predistortion, provided that the respective system is working in its allowable range of pre-amp gain.

The implementation of the DP system in MATLAB and C++ has been proven to be capable of reducing ACPR as high as 50 dB, provided that the signal is applied with a pre-amplifier gain that lies in its allowable range.

References

Lin, C. H., Chen, H. H., Wang, Y. Y., and Chen, J. T. 2006. Dynamically optimum lookup-table spacing for power amplifier predistortion linearization. *IEEE Transactions on Microwave Theory and Techniques* 54:2118–2127.

Varahram, P., Mohammady, S., Hamidon, M. N., Sidek, R. M., and Khatun, S. 2009. Complex gain predistortion in WCDMA power amplifiers with memory effects. Accepted for publication in the *International Arab Journal of Information Technology (IAJIT)*, Jordan, March.

EXPERIMENTAL RESULTS

8.1 Introduction

This chapter presents the experimental results of testing the Mini-Circuit power amplifier using Agilent equipment. VSA software and MATLAB are used for this experiment. The VSA software is used for capturing the samples from the Agilent equipment. There are two main pieces of equipment are used for this testing, a vector signal generator and a vector signal analyzer. The first is used for the digital-to-analog converter (DAC) and upconversion; and the second is used for the analog-to-digital converter (ADC) and downconversion. MATLAB sends the input samples and receives the output samples from the power amplifier (PA) through the vector signal analyzer (VSA) software. These input and output samples can be captured to model the PA and also to apply the complex gain memory predistortion (CGMP) technique and show its effectiveness. The QPSK input signal is applied for this experimental testing. Finally, the experimental results are compared with the simulation results.

8.2 Experimental Setup

For obtaining these results, the Mini-Circuit power amplifier ZVE-8G is applied for this experiment. The reason that a different PA is used for this experiment rather than the PA simulation is because of the availability of this PA for testing. The PA has wideband frequency coverage from 2 GHz to 8 GHz with a 30 dB gain. Here, the PA is working at 2.4 GHz. The input signal which is generated from MATLAB is passed to the Agilent vector signal generator which is upconverted, and then the signal passes it to the power amplifier. The input signal is a quadrature phase shifting key (QPSK) with a sample rate of 1 Mbps and a root rate cosine filter with α that is equal to 0.35. The reason that this signal is selected rather than the previous

Worldwide Interoperability for Microwave Access (WiMAX) or wideband code division multiple access (WCDMA) signal is because the input and output signals should be synchronized together, otherwise the PA characteristics cannot be identified and it becomes asymmetric. The signals that are used in previous chapters for simulation and implementation cannot be modified because they were directly extracted from the Advanced Design System (ADS) and imported to MATLAB, but as mentioned, the signals should be modified in the way that both input and output synchronize together. This is the main reason that these two-carrier WCDMA and WiMAX signals are not used here. The output signal of the PA is connected to the attenuator for bringing down the output power below the input power of the equipment for safety. For a receiving part of the measurement, the 89600S VXI equipment from Agilent is used for which the main task is to downconvert the signal to intermediate frequency (IF). The input signal parameters are shown in Table 8.1. Most of these parameter values are also used in other applications such as digital radio and WiFi (Kenington, 2000).

According to Table 8.1, the carrier frequency is selected to be 2.4 GHz and the input power is 0 dBm. The 2.4 GHz frequency is selected here because of that; this is the ISM (industrial, scientific, and medical) frequency, which is used in most research where there is a measurement (Hong, 2007). Also, the PA can work with this frequency and also the limitations of the equipment that is used and the attenuator that is available is within this range. The symbol rate is 1 Mbps, but another rate can also be applied. Because the modulation is QPSK, then there are four points in constellation and the VSA clock that should capture the received signal should be four times the symbol rate, which is 4 MHz. The number of samples that is sent is 2048 and the roll-off factor for the raised cosine filter is chosen to be 0.35, which is a good value to filter the nonlinearity order (Espinar, 2007).

Table 8.1   Input Signal Parameters

Carrier frequency	2.4 GHz
Input power level	0 dBm
Symbol rate	1 Mbps
VSA clock	4 MHz
Symbol numbers	2048
Roll-off factor (α)	0.35

Here, the VSA software and MATLAB are used for controlling the instruments and sending and receiving the signals. The VSA software can be managed from other applications or software by means of its COM API (Component Object Model Application Programming Interface) language. The main software that can control the VSA are ADS and MATLAB. The latter is used for this work.

More information about VSA software and the command to send data from MATLAB to the equipment can be found in Agilent (2009).

The block diagram of this demonstration is shown in Figure 8.1. As shown in Figure 8.1, the MATLAB and VSA software are running on a PC to control the whole process. The VSA software on a PC will capture the data of the PA. The captured data can then be imported to MATLAB for further analysis.

A photograph of this testing is shown in Figure 8.2. It shows the VSA software that is running to capture the received sample from the PA and also the Agilent equipment for transmitting and receiving the samples.

Synchronization needs to be done before capturing the output data, otherwise the input and output samples will be asymmetric. It is reached with the length capture and search length parameters joined to a synchronization word. This synchronization word can be defined by

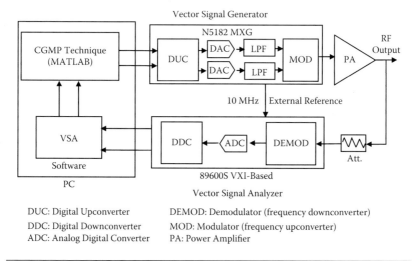

Figure 8.1   Experimental setup for linearization of the PA using the CGMP technique.

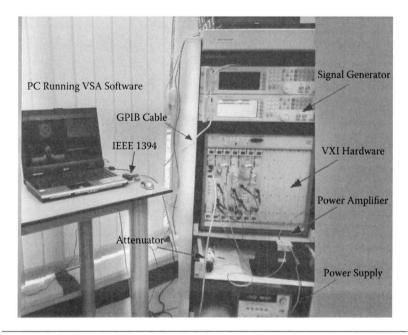

Figure 8.2 Photograph of the experimental setup.

the user and is sent as a pilot message to the signal. Then, the VSA tries to find the pilot message of the signal and, this way, the signals sent and received are completely synchronized. For instance, a 20 symbol pilot message (or synchronization word) is used. The symbols chosen to use here are (00011011000110110001101100011011).

This synchronization word should be added to the input signal, it will then be detected by the VSA software. The result of sending and receiving these samples and importing them into MATLAB is to have a model of the ZVE-8G PA and use this model for testing the capability of the CGMP technique for reduction of the adjacent channel leakage ratio (ACLR). The signal that is sent from MATLAB to the equipment is 2048 samples with QPSK modulation, a 1 Mbps sample rate, and a synchronized word using Standard Command for Programmable Instruments (SCPI). The SCPI defines a standard set of commands to control programmable test and measurement devices. These samples then upconvert using a vector signal generator and pass through the PA. The vector signal generator is connected to the PC with a General Purpose Interface Bus (GPIB) cable. The output of the PA should be attenuated and use a FireWire cable, which should be connected to the other Agilent equipment, the 89600S VXI hardware. This equipment includes several cards. Each of these cards has separate tasks. First, the high-frequency signal is downconverted to IF and then passed to the next card. The other card converts the analog signal to digital with its analog-to-digital converter (ADC). Then the result is sent to the PC using a FireWire (IEEE-1394) cable. The VSA is captured in the samples. Then the sending and receiving samples are used for modeling the PA. By obtaining these samples the characteristics of the PA can be plotted.

The entire test is done offline, meaning that the signal that is received and the transmit signal are not analyzed in real time. Now the CGMP technique is applied to the PA and the amount of reduction is measured. The flowchart for the process is shown in Figure 8.3.

According to this flowchart, the CGMP is applied for the PA and the ACLR reduction, and the error vector magnitude (EVM) is measured. The CGMP is running in MATLAB and is ready to update the LUT values when the transmitting and receiving samples are imported into MATLAB. Each time the samples are imported into MATLAB, the program starts to use the CGMP to multiply them with the input signals. The predistorted samples are sent to the vector signal generator and again they are imported into MATLAB. In fact, this whole process is one iteration. The same procedure happens in any communication system. The initial results for one iteration show an improvement in ACLR. The results will be shown later. For controlling the number of iterations in the CGMP, the error vector should be calculated each time. It is the difference between normalized output and input of the PA. This error vector defines the number of iterations that should be processed until the error reaches the limit. The limit can be specified because the error vector will not be zero for a few iterations and the limit should be specified. Here, it is assumed that the error vector should be less than 0.001. It should be noted that here all the processing is done on a PC rather than by digital signal processing (DSP) or the field programmable gate array (FPGA). For the FPGA interface, another interface to convert I and Q to the Agilent equipment input I and Q is necessary. That device was not available for our testing. The same structure is applied in some good research studies (Hong et al., 2007; Woo et al., 2007).

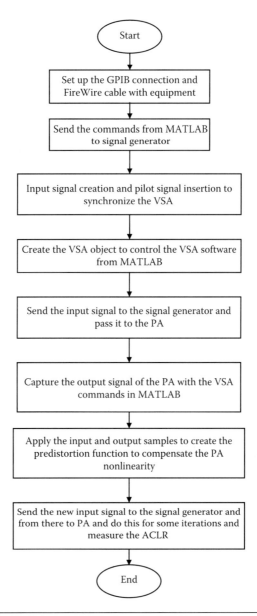

Figure 8.3 Flowchart for experimental testing of the CGMP technique.

8.3 Experimental Results

Here, the experimental results are presented. First, the results of the PA are shown. The AM-AM and AM-PM characteristics of this PA are shown in Figure 8.4. For extracting these figures, the setup that was explained in detail in Figure 8.1 is used. The 2048 samples

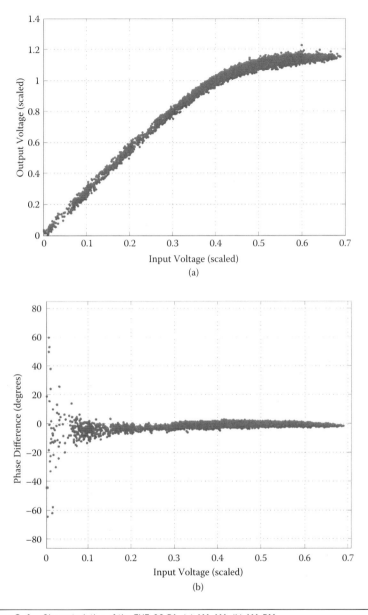

Figure 8.4 Characteristics of the ZVE-8G PA: (a) AM-AM, (b) AM-PM.

that were sent from MATLAB and captured with VSA from the PA output are saved.

Then the same method that was used for modeling the power amplifier MRF1806 in Chapter 3 is also applied to model this power amplifier with memory effects. It can be seen in Figure 8.4 that the scattering of the samples in the AM-AM and AM-PM, which is due to the impact of memory effects as seen in Figure 8.1. The x and y axes are based on normalized input and output voltages and so there is no unit for display.

It is also observed that by increasing the symbol rate, which here is 1 Mbps, the amount of memory is increased, which is expected as the power amplifier shows more memory when the signal bandwidth is increased.

The extracted coefficients of this PA from the input and output samples are:

$a_{10} = 1.524 - 0.211i$; $a_{11} = 0.349 + 0.32i$; $a_{12} = -0.797 - 0.0247i$

$a_{30} = -0.0355 + 0.72i$; $a_{31} = -0.010 - 0.012i$; $a_{32} = -0.0065 + 0.0042i$

$a_{50} = -0.019 - 0.004i$; $a_{51} = 0.009 - 0.019i$; $a_{52} = -0.0069 + 0.013i$

These coefficients are $K = 3$ and $Q = 2$. The memory effect modeling ratio (MEMR) is 0.62, which is more memory than the previous PA.

In Figure 8.5, the result of the VSA software is shown. Panel A is the constellation, the spectrum is panel B, the error magnitude is panel C, and panel D has tables of some parameters like the EVM value. According to Figure 8.5, the constellation of the received signal is distorted and it can also be seen in panel B that the spectrum has nonlinearity and the ACLR is approximately –35 dBm, and the EVM is 3.38%. After importing the data into MATLAB and analyzing them, the predistortion function based on CGMP is generated. Then the new signal is passed through a power amplifier and the result is a power amplifier without distortions. The result of this experiment is shown in Figure 8.6. According to Figure 8.6, the average of the ACLR is –45 dBm, which is a 10 dB reduction when applying the CGMP technique. The result is after five iterations. This means that the predistorted samples are passed through the PA five times.

8.4 Comparison between Simulation and Experimental Results

The EVM is now 1.75%, and the magnitude of the error is significantly reduced. The amount of reduction is significant as this PA is highly linear. It should be considered that when the memory polynomial was

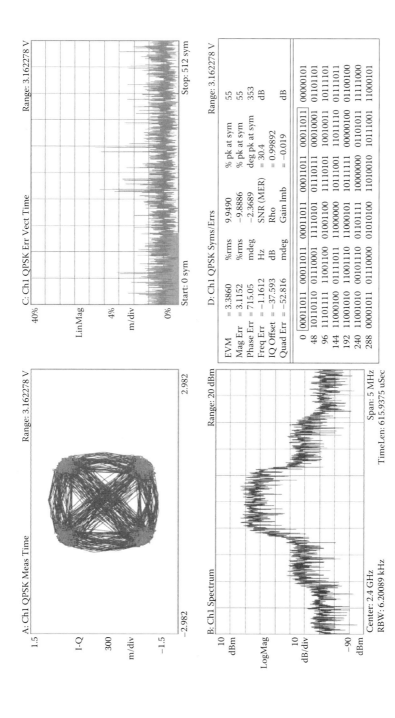

Figure 8.5 VSA software measurements displayed for the ZVE-8G PA without applying CGMP.

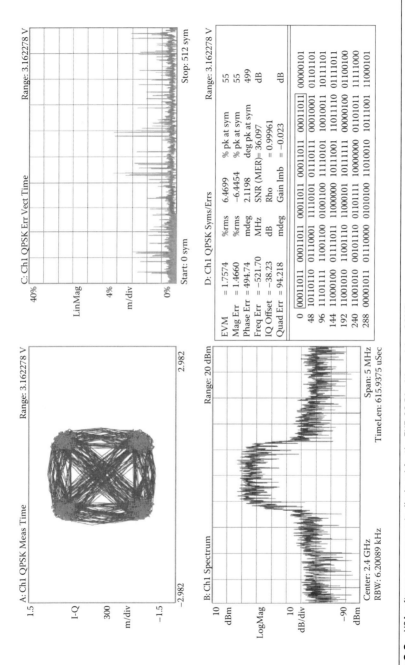

Figure 8.6 VSA software measurements displayed for the ZVE-8G PA with applying CGMP.

used, the reduction in ACLR was −41 dBm and the EVM was 2.1%. The results show the improvement when using the CGMP technique over the memory polynomial technique (Ding et al., 2006).

In Figures 8.7 and 8.8, the power spectral density is shown, which is the same as in Figure 8.5b and Figure 8.6b, but here just the spectrum

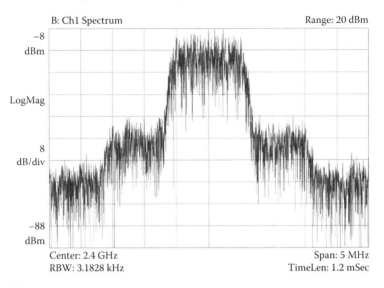

Figure 8.7   Power spectral density of the ZVE-8G power amplifier without CGMP.

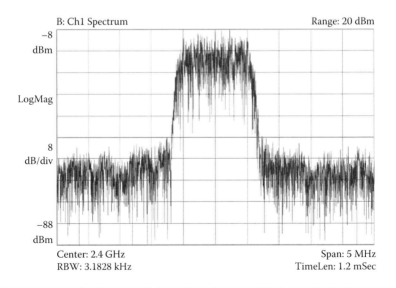

Figure 8.8   Power spectral density of the ZVE-8G power amplifier with applying CGMP after five iterations.

is shown to capture the effect of the CGMP technique in the reduction of ACLR.

Here, the simulation and experimental results are compared. The experimental results as shown are near the simulation results. For doing the comparison, first the simulation should be done with the new PA and the input signal, which are different from the simulations that were done in Chapter 6. The ZVE-8G power amplifier model is applied and the memory polynomial and CGMP technique are applied to this PA with the new input signal. The results of the comparison are shown in Figures 8.9 and 8.10. In Figure 8.9, the power spectral density (PSD) of the experimental results is compared with the simulation. Figure 8.9a shows the PSD of the power amplifier without predistortion and it can be seen that the ACLR is around −35 dB. After applying the CGMP, the experimental result is shown in Figure 8.9b; here the ACLR is around −45 dB. The simulation result is also added and is shown in Figure 8.9c, the ACLR is now almost −50 dB. It is obvious that there is a difference between experimental and simulation results, and the reason is the effects of DAC, mixer, and upconversion are not considered in the simulations and those effects along with effects of the power

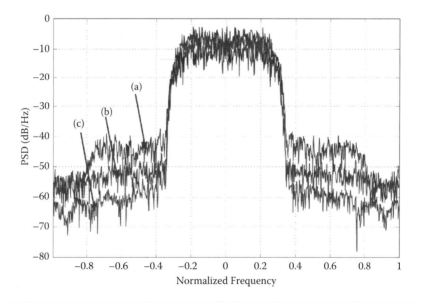

Figure 8.9 Comparison of the PSD for the ZVE-8G power amplifier: (a) without CGMP; (b) experimental PSD with CGMP and after five iterations; and (c) simulation PSD with CGMP after five iterations.

amplifier memory, which here is modeled for a memory length of 2, can cause this difference.

In Figure 8.10, the comparison of the power spectral density between the memory polynomial technique and the CGMP technique is shown. There is almost 3 dB to 5 dB more reduction when the CGMP technique is applied. Figure 8.10 focuses on the PSD of the left side

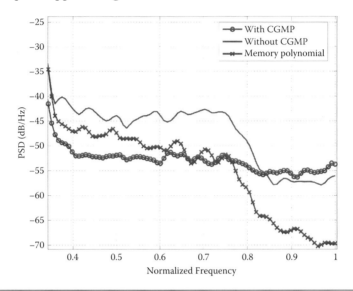

Figure 8.10 Comparison of the PSD for the ZVE-8G power amplifier: (a) without CGMP; (b) with memory polynomial method with $Q = 2$; and (c) with CGMP technique after five iterations.

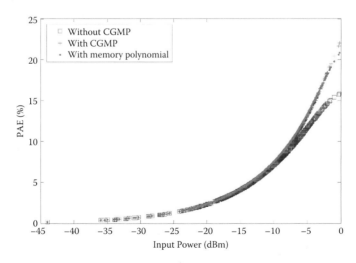

Figure 8.11 PAE of the ZVE-8G power amplifier with and without CGMP and with the memory polynomial method.

to distinguish the difference. Both the left side and right side of the PSD have equal ACLR. In Figure 8.11, the power-added efficiency (PAE) of this power amplifier is shown. By applying the CGMP technique, around 7% improvement in efficiency of the power amplifier is achieved when the input power is 0 dBm. The efficiency improvement is even not so significant but is acceptable because of the type of the power amplifier that is highly linear. By considering a highly nonlinear PA, the efficiency will be enhanced further. This enhancement is without the loss of linearity. The value is less when the memory polynomial method is applied.

8.5 Summary

In this chapter, the experimental results with Agilent equipment and using the Mini-Circuit PA is presented. The VSA software is used for capturing the received samples from the equipment. MATLAB controlled the whole testing. It sends the samples with SCPI commands and after the samples are received from the VSA, then the input and output samples are analyzed and the CGMP technique can be applied to the input samples. The modified samples are then transmitted to the PA to reduce the nonlinearity of the PA. The QPSK signal is applied with a 1 Mbps sample rate for the input signal. The reduction of 10 dB in ACLR is achieved after applying the CGMP technique. Also, the experimental results are compared with the simulation results and memory polynomial technique to validate the proposed technique.

References

Cripps, S. C. 2006. *RF Power Amplifiers for Wireless Communications*. Norwood, MA: Artech House.

Davis, M. 2002. A software linearization method for enhancing noise loaded performance of amplifiers. American Institute of Aeronautics and Astronautics.

Ercegovac, M. D, and Muller, J. M. 2003. Complex division with prescaling of operands. IEEE International Conference on Application-Specific Systems, Architectures and Processors, pp. 304–314.

Espinar, J. M. 2007. Digital predistortion by using GPIB-controlled instrumentation. Master's thesis, Universititat Politechnica de Catalunya.

Hong, S., Woo, Y. Y., Kim, J., Cha, J., Kim, I., Moon, J., Yi, J., and Kim, B. 2007. Weighted polynomial digital predistortion for low memory effect Doherty power amplifier. *IEEE Transactions for Microwave Theory Technology* 55(5):925–931, May.

Hyunsuk, M., and Sedaghat, R. 2006. FPGA-based adaptive digital predistortion for radio-over-fiber links. *Microprocessors and Microsystems* 30(3):145–154.

Kenington, P. B. 2000. *High-Linearity RF Amplifier Design*. Norwood, MA: Artech House.

Woo, Y. Y., Kim, J., Hong, S., Kim, I., Moon, J., Yi, J., and Kim, B. 2007. Adaptive digital feedback predistortion technique for linearizing power amplifiers. *IEEE Transactions for Microwave Theory Technology* 55(5):932–940.

Appendix A: Complex Baseband Representation of Band-Pass Signals

In this book, analog parts of communication systems like digital-to-analog converters and upconversion units are not considered because of some of the effects they have. Some researchers have demonstrated these effects as analog imperfections, and studied and then modeled them (Cavers 1997; Ding, 2004). Here, we will show how the band-pass signal can be modeled in baseband without losing information, which will prove that there is no need to convert the baseband signal to RF for complete modeling. In all the simulations, we use the baseband process rather than convert it to RF.

The first step in the development of a complex baseband representation is to define a band-pass signal. A band-pass signal, $x_c(t)$, is a signal where the one-sided energy spectrum is both:

- Centered at a nonzero frequency, f_c, and
- Does not extend in frequency to zero (DC)

Typically, the bandwidth of a signal is expressed by B_T Hertz. This means that the band-pass signal should satisfy the constraint $\dfrac{B_T}{2} \leq f_c$. Figure A.1 shows a typical band-pass spectrum.

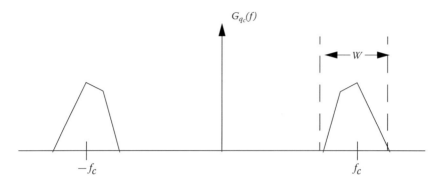

Figure A.1 Energy spectrum of a band-pass signal. (From Fitz, M. P. 2007. *Fundamentals of Communications Systems*. New York: McGraw-Hill.)

A band-pass signal has a representation of

$$x_c(t) = x_I(t)\sqrt{2}\cos(2pf_ct) - x_Q(t)\sqrt{2}\sin(2pf_ct)$$

$$= x_A(t)\sqrt{2}\cos(2pf_ct + x_P(t)) \tag{A.1}$$

where f_c is the carrier frequency with $f_c - \dfrac{B_T}{2} \leq f_c \leq f_c + \dfrac{B_T}{2}$. The signal $x_I(t)$ in Equation (A.1) is normally referred to as the in-phase (I) component of the signal and the signal $x_Q(t)$ is normally referred to as the quadrature (Q) component of the band-pass signal. The transformations between the two representations are given by

$$x_A(t) = \sqrt{x_I(t)^2 + x_Q(t)^2}$$

$$x_P(t) = \tan^{-1}\left[x_Q(t), x_I(t)\right] \tag{A.2}$$

$$x_I(t) = x_A(t)\cos(x_P(t))$$

$$x_Q(t) = x_A(t)\sin(x_P(t)) \tag{A.3}$$

Note that the tangent function in Equation (A.2) has a range of $(-\pi, \pi)$ (i.e., both the sign $x_I(t)$ and $x_Q(t)$ and the ratio of $x_I(t)$ and $x_Q(t)$ are needed to evaluate the function).

A complex signal also can be represented as:

$$x_z(t) = x_I(t) + jx_Q(t) = x_A(t)\,exp\left[jx_P(t)\right] \tag{A.4}$$

The original band-pass signal can be obtained from the complex envelope by

$$x_c(t) = \sqrt{2}R\left[x_z(t)\exp[j2\pi f_c t]\right] \tag{A.5}$$

Since the complex exponential only determines the center frequency, the complex signal $x_z(t)$ contains all the information in $x_z(t)$. By using this complex baseband representation of band-pass signals, the simulations of communication systems is simpler. The next thing to consider is the methods used to translate between a band-pass signal and a complex envelope signal. Basically, a band-pass signal is generated from its I and Q components as shown in Equation (A.4). Similarly, the complex envelope signal can be shown as below:

$$x_c(t)\sqrt{2}\cos(2\pi f_c t) = x_I(t) + x_I(t)\cos(4\pi f_c t) - x_Q(t)\sin(4\pi f_c t) \tag{A.6}$$

$$x_c(t)\sqrt{2}\cos(2\pi f_c t) = -x_Q(t) + x_Q(t)\cos(4\pi f_c t) + x_I(t)\sin(4\pi f_c t) \tag{A.7}$$

Figure A.2 shows these transformations where the low-pass filters remove the $2f_c$ terms in Equation (A.6). In Figure A.2, note that the boxes with $\pi/2$ are phase shifters (i.e., $\cos(\theta - \pi/2) = \sin(\theta)$), which are typically implemented with delay elements. The structure in Figure A.2 is fundamental to the study of all modulation techniques. We want to derive the spectral representation of the complex baseband signal, $x_z(t)$, and compare it to the spectral representation of the band-pass signal, $x_c(t)$.

Complex Baseband to Band-Pass Conversion   Band-Pass to Complex Band-Pass Conversion

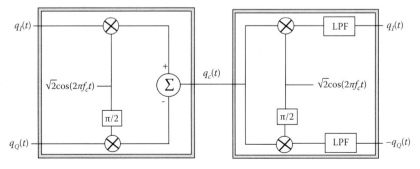

Figure A.2 Schemes for converting between complex baseband and band-pass representations. Note that the LPF simply removes the double frequency term associated with the downconversion. (From Fitz, M. P. 2007. *Fundamentals of Communications Systems.* New York: McGraw-Hill.)

Assume that x_z (t) is an energy signal, the Fourier transform of x_z (t) is

$$X_z(f) = X_I(f) + jX_Q(f) \qquad (A.8)$$

where $X_I(f)$ and $X_Q(f)$ are the Fourier transforms of $x_I(t)$ and $x_Q(t)$, respectively, and the energy spectrum is given by

$$G_{x_z}(f) = G_{x_I}(f) + G_{x_Q}(f) + 2\mathfrak{I}\left[X_I(f)X_Q^*(f)\right] \qquad (A.9)$$

where $G_{x_I}(f)$ and $G_{x_Q}(f)$ are the energy spectrum of $x_I(t)$ and $x_Q(t)$, respectively.

Note that $x_I(t)$ and $x_Q(t)$ are both real signals so that $X_I(f)$ and $X_Q(f)$ are Hermitian symmetric functions of frequency, it is then straightforward to show

$$\begin{aligned} X_z(-f) &= X_I^*(f) + jX_Q^*(f) \\ X_z^*(-f) &= X_I(f) - jX_Q(f) \end{aligned} \qquad (A.10)$$

This leads directly to

$$\begin{aligned} X_I(f) &= \frac{X_z(f) + X_z^*(-f)}{2} \\ X_Q(f) &= \frac{X_z(f) - X_z^*(-f)}{j2} \end{aligned} \qquad (A.11)$$

Since $x_z(t)$ is a complex signal, in general, the energy spectrum, $G_{xz}(t)$, has none of the usual properties of real signal spectra. The same method can be used to create the spectral characteristics of the band-pass signal. The Fourier transform of the band-pass signal, $x_c(t)$, is expressed as

$$\begin{aligned} X_c(f) &= \left[\frac{1}{\sqrt{2}}X_I(f - f_c) + \frac{1}{\sqrt{2}}X_I(f + f_c)\right] \\ &\quad -\left[\frac{1}{j\sqrt{2}}X_Q(f - f_c) + \frac{1}{j\sqrt{2}}X_Q(f + f_c)\right] \end{aligned} \qquad (A.12)$$

and

$$X_c(f) = \frac{1}{\sqrt{2}} X_z(f - f_c) + \frac{1}{\sqrt{2}} X_z^*(-f - f_c) \qquad (A.13)$$

This is a very fundamental result. Equation (A.13) states that the Fourier transform of a band-pass signal is simply derived from the spectrum of the complex envelope. For positive values of f, $X_c(f)$ is obtained by translating $X_z(f)$ to f_c and scaling the amplitude by $\frac{1}{\sqrt{2}}$. For the negative values of f, $X_c(f)$ is obtained by flipping $X_z(f)$ around the origin, taking the complex conjugate, translating the result to $-f_c$, and scaling the amplitude by $\frac{1}{\sqrt{2}}$. This also demonstrates that if $X_c(f)$ only takes values when the absolute value of f is in $(f_c - B_T, f_c + B_T)$, then $X_z(f)$ only takes values in $(-B_T, B_T)$. The energy spectrum of $X_c(f)$ can also be expressed in terms of the energy spectrum of $x_z(t)$ as

$$G_{x_c}(f) = \frac{1}{2} G_{x_z}(f - f_c) + \frac{1}{2} G_{x_z}(-f - f_c) \qquad (A.14)$$

Since

$$E_{x_c} = \int_{-\infty}^{\infty} G_{x_c}(f) df = \int_{-\infty}^{\infty} G(f) df \qquad (A.15)$$

This will ensure that the energy of the complex envelope is identical to the energy of the band-pass signal. Considering these results, the spectrum of the complex envelope of the signal shown in Figure A.1 will have a form shown in Figure A.3 when $f_c = f_C$. Other values of f_c would produce a different but equivalent complex envelope representation.

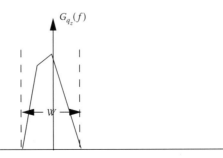

Figure A.3 The complex envelope energy spectrum of the band-pass signal. (From Fitz, M. P. 2007. *Fundamentals of Communications Systems.* New York: McGraw-Hill.)

References

Cavers, J. K. 1997. The effect of quadrature modulator and demodulator errors on adaptive digital predistorters for amplifier linearization. *IEEE Transactions on Vehicular Technology* 46(2):456–466.

Ding, L., Zhou, G., Morgan, D., Ma, Z., Kenny, J., Kim, J., and Giardina, C. 2004. A robust digital baseband predistorter constructed using memory polynomials. *IEEE Transactions on Communications* 52(1), January.

Fitz, M. P. 2007. *Fundamentals of Communications Systems.* New York: McGraw-Hill.

Appendix B

B.1 Power Amplifier Model with Memory Effects

```
%%%%%%%%%%%%%%%%%%%%%%%%%%%%%%%%%%%%%%%%%%%%%%%%%%%%%%%%%%%%
%%%%%%%%%%%%%%%%%%%%%%%%%%%%%%%%%%%%%
%       LINEARIZATION WITH NONADAPTIVE PREDISTORTION
%       STEEPEST DESCENT WITH SCALE & ROTATE PARAMETERS
USED TO CONSTRUCT LUT
%                         ____
%              vi(n) |    |   | y(n)  |    |   vo(n)
%             ------>| PD |--------->| PA |------->
%                |   |____|          |____|   |
%                |   \/____/         \____/   |
%                |    |            saleh-model |
%%%%%%%%%%%%%%%%%%%%%%%%%%%%%%%%%%%%%%%%%%%%%%%%%%%%%%%%%%%%
%%%%%%%%%%%%%%%%%%%%%%%%%%%%%%%%%%
close all;
clear;

PA_Model=0;  %  0= Static Model,
               1= Dynamic Model   (step variation),
               2= Dynamic Model   (memory effect)
numtones=2;
f1=100;
f2=112;
f3=14;
f4=16;
```

```
f5=18;
f6=20;


g1=3;
g2=2;
g3=-10;
g4=-10;
g5=-10;
g6=-10;
if numtones == 1
        f2=10;
        g2=0;
end
f1=f1*1E6;
f2=f2*1E6;
f3=f3*1E6;
f4=f4*1E6;
f5=f5*1E6;
f6=f6*1E6;


Fs=125E6;
BufferSize=512;
fo=Fs/BufferSize;
n1=round(f1/fo);
n2=round(f2/fo);
n3=round(f3/fo);
n4=round(f4/fo);
n5=round(f5/fo);
n6=round(f6/fo);
n=0:(BufferSize-1);
N1=BufferSize/n1;
N2=BufferSize/n2;
N3=BufferSize/n3;
N4=BufferSize/n4;
N5=BufferSize/n5;
N6=BufferSize/n6;


NumBits=16;
ScaleFactor=2.^(NumBits-1)-1;


g1=10.^(g1/20);
g2=10.^(g2/20);
g3=10.^(g3/20);
g4=10.^(g4/20);
```

```
g5=10.^(g5/20);
g6=10.^(g6/20);

if numtones == 6
          y =     ( ( 1/2 * ( g1 * exp(j*2*pi*n/N1) +
                  g2 * exp(j*2*pi*n/N2) + g3 *
                  exp(j*2*pi*n/N3) + g4 *
                  exp(j*2*pi*n/N4) + g5 *
                  exp(j*2*pi*n/N5) + g6 *
                  exp(j*2*pi*n/N6)) ) );

elseif  numtones == 3
          y =     (( 1/2 * ( g1 * exp(j*2*pi*n/N1) +
                  g2 * exp(j*2*pi*n/N2) + g3 *
                  exp(j*2*pi*n/N3)) ) );

elseif numtones == 2
          y =     round( ScaleFactor* ( 1/2 * ( g1 *
                  exp(j*2*pi*n/N1) + g2 *
                  exp(j*2*pi*n/N2) ) ) );
elseif numtones == 1
          y =     round( ScaleFactor * g1 *
                  exp(j*2*pi*n/N1) );
end
y1=y;
%Inp=y;
%Inp=Inp./max(abs(Inp));
%Inp=.7.*Inp.';
%figure;
%plot(linspace(-Fs/2, Fs/2, length(Inp)), 20*log10(abs
  (fftshift(fft(Inp)))));
      %%%%%%%%%%%%%%%%%%%%%%%%%%%%%%%%%%%%%%%%%%%%%%%%
      %%             Random bit Generator          %%
      %%%%%%%%%%%%%%%%%%%%%%%%%%%%%%%%%%%%%%%%%%%%%%%%
data=dlmread('IQ.txt');
I=data(:,1);
Q=data(:,2);
Inp=0.8.*(I+i*Q);
%Inp=Inp./max(abs(Inp));
%Inp=((randint(1,N)-.5)*2+j*(randint(1,N)-.5)*2)/
      2^.10;
%Random bir Generator(QPSK)
%Inp=Inp.*exp(j.*2.*pi.*N/N1);
Fs=1250E06;
f=100E06;
```

```
%Inp=decimate(Inp,1);
Inp=decimate(Inp,2);
%Inp=Inp.*exp(j.*2.*pi./10)+Inp.*exp(-j.*2.*pi./1000);
%figure;
%plot(20*log10(abs(fft(Inp))))
%hold on;
%plot(linspace(-Fs/2, Fs/2, length(Inp1)),
    20*log10(abs(fftshift(fft(Inp1))))),'r')
%Inp=rcosflt(Inp,1, 8,'sqrt', 0.4,4);
%Inp=real(Inp).*cos(j*2.*pi.*f1./Fs)+i*imag(Inp).
    *sin(j*2.*pi.*f1./Fs);
%Inp=in;

%Inp=Inp.*exp(j*2*pi.*f./Fs);
%I=real(Inp);
%Q=imag(Inp);
%data=zeros(2,r);
%data=data';
%data(:,1)=I;
%data(:,2)=Q;
%Inp=Inp(80:end-80);
%Inp=[1/Res:1/Res:.5];    %Random bit Generator(QPSK)
%Inp=interp(Inp,1);
Out=Saleh_model_modified(Inp,PA_Model);
%x=Inp;
%Y=(Out);

load IQ-2.mat;
load X-2.mat;
%Y=Y(1:4096);
x=x1;
Inp=x;
r=length(Inp);
m=2;
n=3;

U=ones(20,r);
a=zeros(1,n*(m+1));
a=a.';
b=ones(20,r);
Hq=zeros(n,r-m);
Hq=Hq.';
H=zeros(r-m,n*(m+1));
%d=zeros(1,r);
d1=ones(1,r);
```

```
d1=d1.';
%d=d.';
H=H;
b=b;
l=0;
%q=0;
rmse=zeros(1,r);
rmse=rmse.';
rmse0=zeros(1,r);
rmse0=rmse0.';
    for q=0:m
        for k=1:n
            for s=1:r-q
                d=x(q+1:end);
                d1(s)=d(s);
            h=d1.*(abs(d1)).^((k-1));
            h=h(1:end-m);
        end;
        Hq(:,k)=h;
        end;
        ct=(n).*q;
        for j=1:n
        H(:,ct+j)=Hq(:,j);
        end;
        d1=ones(1,r);
d1=d1.';

    end;
    for p=1:1

    Y=Y(1:end-m);
    a=pinv(H)*Y;
    d1=ones(1,r-m);
        d1=d1.';

Inp=Inp(1:end-m);
  y=zeros(1,r-m);
  y=y.';

  for f=0:0
      for k=1:n
          for v=1:r-m
              x=Inp;
              d=x(f+1:end);
              d1(v)=d(v);
      c=(n).*f;
```

```
        end;
          y=y+d1.*(abs(d1)).^(k-1)*a(k,:);
        end;
    E=Y-y;
  for l=1:r-m
  rmse0=rmse0+E(l,:).^2;
end;
rmse0=sqrt(rmse0);

  end;
  d1=ones(1,r-m);
      d1=d1.';
    for f=1:m
      for k=1:n
        for v=1:r-m-f
            x=Inp;
            d=x(f+1:end);
            d1(v)=d(v);
        c=(n).*f;

      end;
        y=y+d1.*(abs(d1)).^(k-1)*a(k+c,:);
      end;
    d1=ones(1,r-m);
        d1=d1.';
  end;

  E=Y-y;
  Y=y;
  for l=1:r-m
  rmse=rmse+E(l,:).^2;
end;
rmse=sqrt(rmse);
MEMR=1-rmse./rmse0;
end;
  %Out=Saleh_model_modified(Zpd,PA_Model);
    Inp=Inp(1:end);
    Out=Out(1:end-m);
  y=y;
figure;
plot(abs(Inp),abs(Out),'r.');
hold on;
plot(abs(Inp),abs(y),'.');
figure;
psd(y);
```

```
hold on;
psd(Out);
figure;
plot(abs(MEMR))
```

B.1.1 *Mini-Circuit Power Amplifier Model without Memory*

```
function Out=PAZVE(Inp)
load X-5.mat;
load IQ-5.mat;
%Y=Y(1:4096);
%x1=interp(x1,2);
%Y=interp(Y,2);
%Inp=0.8.*x1;
u=abs(Inp);
Pout=abs(Y);
Pin=abs(x1);
Phase=pi./180.*phase(Y);
AMAMco=polyfit(Pin,Pout,12);
AMPMco=polyfit(Pin,Phase,12);

AMAM=polyval(AMAMco,u);
%AMAM=AMAM./max(AMAM);
AMPM=polyval(AMPMco,u);
Phi=unwrap(AMPM+angle(Inp));
Out=AMAM.*exp(j*Phi);
%figure;
%plot(u,abs(Out),'.');
%figure;
%plot(u,unwrap(phase(Out)-phase(Inp)),'.');
%figure;
%psd(Out);
```

B.2 Mini-Circuit Power Amplifier Model with Memory

```
function y=PA(Inp)

c=zeros(1,9);
r=length(Inp);

c(1,1)=1.52448089614002 - 0.211207277450019i
c(1,2)=0.349763648695264 + 0.322758521656134i
c(1,3)=-0.797887065296552 - 0.0247554777024818i
c(1,4)=0.0355866125933398 + 0.0728400541776913i
c(1,5)=-0.0101908426014843 - 0.0125685167203472i
```

```
c(1,6)=-0.00653529617894296 + 0.00427217964806115i
c(1,7)=-0.0192398088333449 - 0.00469372132606898i
c(1,8)=0.00945176787508824 - 0.0195446838509838i
c(1,9)=-0.00691477057584259 + 0.0130320964377561i

n=3;
m=2;
r=length(Inp);
c=c.';
a=zeros(1,n);
a=a.';
y=0;
Hq=zeros(r,1);
  d1=ones(1,r);
        d1=d1.';
      for q=0:m
        Hq=zeros(r,1);
        for k=1:n
                for s=(1+q):r
                    for j=1:n
                        ct=n*q;
                        a(j,:)=c(j+ct,:);
                    end;
                    x=Inp;

                    Hq(s,1)=x(s-q);

                end;
                Zp=a(k,:).*Hq.*(abs(Hq)).^(2*(k-1));

        y=y+Zp;
      end;

        a=zeros(1,n);
        a=a.';

      end;
y=y;

  %y=y(1:end-2);
```

B.3 Predistortion Filter Based on Memory Polynomial

```
close all;
clear;
```

```
PA_Model=0;  %        0= Static Model,
                      1= Dynamic Model (step variation),
                      2= Dynamic Model (memory effect)
     %%%%%%%%%%%%%%%%%%%%%%%%%%%%%%%%%%%%%%%%%%%%%%%%
     %%              Random bit Generator           %%
     %%%%%%%%%%%%%%%%%%%%%%%%%%%%%%%%%%%%%%%%%%%%%%%%

%Inp=Inp./max(abs(Inp));
%Inp=((randint(1,N)-.5)*2+j*
    (randint(1,N)-.5)*2)/2^.10;
%Random bir Generator(QPSK)
%Inp=Inp.*exp(j.*2.*pi.*N/N1);
Fs=1250E06;
f=100E06;
%Inp=decimate(Inp,1);
%Inp=decimate(Inp,1);
%Inp=Inp.*exp(j.*2.*pi./10)+Inp.*exp(-j.*2.*pi./1000);
%figure;
%plot(20*log10(abs(fft(Inp))))
%hold on;
%plot(linspace(-Fs/2, Fs/2, length(Inp1)),
    20*log10(abs(fftshift(fft(Inp1)))),'r')
%Inp=rcosflt(Inp,1, 8,'sqrt', 0.4,4);
%Inp=real(Inp).*cos(j*2.*pi.*f1./Fs)+i*imag(Inp).
    *sin(j*2.*pi.*f1./Fs);
%Inp=in;

load xIyI_samples.mat

seqLen = length(xI);
Ax_sorted = sort(abs(xI));

yI_orig = yI;
yI_new = align_signals(yI, xI);
yI = yI_new;
gain = yI_new./xI;
Inp=(xI);
Inp=Inp(1:1024*4);

Inp=Inp.';
 %Inp = Inp / sqrt(mean(abs(Inp).^2));
 %Out = Out / sqrt(mean(abs(Out).^2));
%Out=yI.';
Out=gain(1:1024*4).';
%Inp=(Iin+i*Qin);
```

```
%Inp=Inp./(max(abs(Inp)));
%Out=(Iout+i*Qout);
%Out=Out./(max(abs(Out)));
%Out=Saleh_model_modified(Inp,PA_Model);
%Inp=Inp(1:end-2);
x=Inp;
r=length(Inp);
Y=(Out);
n=3;
m=2;


U=ones(20,r);
a=zeros(1,n*(m+1));
a=a.';
b=ones(20,r);
Hq=zeros(r,n);
%Hq=Hq.';
H=zeros(r,n*(m+1));
%d=zeros(1,r);
d1=ones(1,r);
d1=d1.';
%d=d.';
H=H;
b=b;
l=0;
%q=0;
rmse=zeros(1,r);
rmse=rmse.';
rmse0=zeros(1,r);
rmse0=rmse0.';
    for q=0:m
        Hq=zeros(r,n);
        for k=1:n
            for s = (q+1):r
                d1=x(s-q);

            Hq(s,k)=d1.*(abs(d1)).^(2*(k-1));

            end;
        end;
        ct=(n).*q;
        for j=1:n
        H(:,ct+j)=Hq(:,j);
        end;
```

```
      end;
   for p=1:1

   %Y=Y(1:end);
   a=pinv(H)*Y;
    d1=ones(1,r-m);
        d1=d1.';

Inp=Inp(1:end-m);
 y=zeros(1,r-m);
 y=y.';

 for f=0:0
     for k=1:n
        for v=1:r-m
                x=Inp;
                d=x(f+1:end);
                d1(v)=d(v);
                c=(n).*f;

        end;
           y=y+d1.*(abs(d1)).^(2*(k-1))*a(k,:);
        end;
   E=Y-y;
 for l=1:r-m
 rmse0=rmse0+E(l,:).^2;
end;
rmse0=sqrt(rmse0);

   end;
 d1=ones(1,r-m);
       d1=d1.';
   for f=1:m
      for k=1:n
        for v=1:r-m-f
                x=Inp;
                 d=x(f+1:end);
                 d1(v)=d(v);
            c=(n).*f;

    end;
      y=y+d1.*(abs(d1)).^(2*(k-1))*a(k+c,:);
   end;
  d1=ones(1,r-m);
      d1=d1.';
 end;
```

```
E=Y-y;
Y=y;
for l=1:r-m
rmse=rmse+E(l,:).^2;
end;
rmse=sqrt(rmse);
MEMR=1-rmse./rmse0;
end;
%Out=Saleh_model_modified(Zpd,PA_Model);
    Inp=Inp(1:end-2);
    Out=Out(1:end-m-2);
y=y(1:end-2);
figure;
%plot(abs(Inp),abs(Out),'r.');
%hold on;
plot(abs(Inp),abs(y),'.');
figure;
psd(y);
hold on;
psd(Inp);
```

B.4 Predistortion Filter Based on Complex Gain

```
close all;
clear;
iteration=3;

Res=1024;
%%%%%%%%%%%%%%%%%%%%%%%%%%%%%%%%%%%%%%%%%%%%%%%
            %%          LUT Generation          %%
PA_Model=7;     %   0= Static Model,
                1= Dynamic Model (step variation),
                2= Dynamic Model (memory effect)

%load F.mat

            %%%%%%%%%%%%%%%%%%%%%%%%%%%%%%%%%%%%%%
            %%      Random bit Generator      %%
            %%%%%%%%%%%%%%%%%%%%%%%%%%%%%%%%%%%%%%
%Res=2^10;     Gmax                 %LUT Resolution
%N=1024/8;               % number of random input
%Inp=0.988*((randint(1,N)-.5)*2+j*(randint(1,N)-.5)*2)/
    2^.5;
```

```
%Random bit Generator(QPSK)
%Inp=(rcosflt(Inp,1, 8,'sqrt', 0.35,4))';
%Inp=Inp(33:end-32);
%Inp=Interp(Inp,8);

load X-2.mat;
Inp=x1;
r=length(Inp);
Inp=Inp.'; %Passing Random Input through Saleh Model
   for Plotting primary power Amp.

data=dlmread('IQCDMA1.txt');
I=data(:,1);
Q=data(:,2);
%Inp=0.9.*(I+i*Q);
%Inp=Inp(1:2000);

Outm=Saleh_model_modified(Inp,PA_Model); %Passing
   Input Data through Saleh model for Linearizing
%Out0=PAS0(Inp);

                      %find maximum gain of PA
Gmax=1.0215;
Gmax=2.25;
%Gmax=2.1587;
%Gmax=4.73;
F=ones(2000,1);

S=ones(1024,1);
R=zeros(1024,1);

      %%%%%%%%%%%%%%%%%%%%%%%%%%%%%%%%%%%%%%%%%%%%%%
      %            Index for Lookup Table          %
      %%%%%%%%%%%%%%%%%%%%%%%%%%%%%%%%%%%%%%%%%%%%%%

%%%%%%%%%%%%%%%%%%%%%%%%%%%%%%%%%%%%%%%%%%%%%%%%%%%%%%%%%%
%%%%%%%%%%%%%%%%%%%%%%%%%%%%%%%%%%%%%%%%%%%%%%%%%%%%%%%%%%
%for k=1:length(Inp)
%    Indx(k)=round(1024*abs(Inp(k)))+1;
%Scale Inputs with LUT Resolution and then index those
%end
%Indx=Indx(1:1024);
%Indx=Indx;
%F=F(Indx);
```

```
%S=S(Indx);%
%R=R(Indx);
%%%%%%%%%%%%%%%%%%%%%%%%%%%%%%%%%%%%%%%%%%%%%%%%%%%%%%%
%%%%%%%%%%%%%%%%%%%%%%%%%%%%%%%%%%%%%%%%%%%%

        %%%%%%%%%%%%%%%%%%%%%%%%%%%%%%%%%%%%%%%%%%%%%%
        %           ADAPTATION ALGORITHM            %
        %           Steepest Descent Algorithm      %
        %%%%%%%%%%%%%%%%%%%%%%%%%%%%%%%%%%%%%%%%%%%%

%%%%%%%%%%%%%%%%%%%%%%%%%%%%%%%%%%%%%%%%%%%%%%%%%%%%%%%%%%%
%%%%%%%%%%%%%%%%%%%%%%%%%%%%%%%%%%%%%%%%%%%%%%
Inp=Inp.';
for i = 1:iteration

        Inpp=F.*Inp;                %Inpp is Predistorted Input

        Outpd=Saleh_model_modified(Inpp,PA_Model);

        Verr=(Outpd)-(Gmax.*(Inp));      %Error(Amplitude)
                                              between output
                                              and input
        %Out0=PAS0(Inp);
        %Out0=interp(Out0,8);
        %Verr=interp(Verr,8);
        alfa=0.6./(Gmax.*Inp);

        F=F-alfa.*Verr;      % R(Rotate) is gain Phase of
                                    Power Amplifier

end;
%%%%%%%%%%%%%%%%%%%%%%%%%%%%%%%%%%%%%%%%%%%%%%%%%%%%%%%%%%%
%%%%%%%%%%%%%%%%%%%%%%%%%%%%%%%%%%%%%%%%%%%

%%%%%%%%%%%%%%%%%%%%%%%%%%%%%%%%%%%%%%%%%%%%%%%%
        %    Data is Predistorted and now inserted
        %    to PA
%%%%%%%%%%%%%%%%%%%%%%%%%%%%%%%%%%%%%%%%%%%%%%%%

%Inp=decimate(Inp,8);
%Outpd=decimate(Outpd,8);
%Outm=decimate(Outm,8);
Fs=1;
h = spectrum.welch('hann',[512]);
figure;
psd(h,Outpd./Gmax,'Fs',Fs,'CenterDC',true)
```

```
%psd(Out./GG);
hold on;
figure;
psd(h,Outm./Gmax,'Fs',Fs,'CenterDC',true)
%psd(Inp);


hold on;
figure;
psd(h,Inp,'Fs',Fs,'CenterDC',true)
```

B.5 Program for Experimental Measurement

```
clear;
io = agt_newconnection('gpib',0,20);
[status, status_description,query_result] =
   agt_query(io,'*idn?');
if (status < 0) return; end
agt_sendcommand(io,'SOUR:RAD:ARB:STAT Off')
agt_sendcommand (io, 'FREQuency 2400000000')
agt_sendcommand (io, 'POWer -7')
[ status, status_description ] = agt_sendcommand
    ( io, 'OUTPut:STATe ON' );
[ status, status_description ] = agt_sendcommand
    ( io, 'POW:ALC:STAT Off');


PA_Model=11;
% INPUT SIGNAL
load X-5.mat;
load F.mat;
x1=F.*x1;


% and a sine waveform in Q.
%
points = 1000; % Number of points in the waveform
cycles = 101; % Determines the frequency offset from
   the carrier
phaseInc = 2*pi*cycles/points;
phase = phaseInc * [0:points-1];
Iwave = cos(phase);
Qwave = sin(phase);
% 2.) Save waveform in internal format
**********************************
% Convert the I and Q data into the internal arb format
```

```
% The internal arb format is a single waveform
    containing interleaved IQ
% data. The I/Q data is signed short integers
    (16 bits).
% The data has values scaled between +-32767 where
% DAC Value Description
% 32767 Maximum positive value of the DAC
% 0 Zero out of the DAC
% -32767 Maximum negative value of the DAC
% The internal arb expects the data bytes to be in Big
    Endian format.
% This is opposite of how short integers are saved on
    a PC (Little Endian).
% For this reason the data bytes are swapped before
    being saved.
% Interleave the IQ data
waveform(1:2:2*points) = Iwave;
waveform(2:2:2*points) = Qwave;
%[Iwave;Qwave];
%waveform = waveform(:)';
% Normalize the data between +-1
waveform = waveform / max(abs(waveform)); % Watch out
    for divide by zero.
% Scale to use full range of the DAC
waveform = round(waveform * 32767); % Data is now
    effectively signed short integer values
% waveform = round(waveform *
    (32767 / max(abs(waveform)))); % More efficient
    than previous twosteps!
% PRESERVE THE BIT PATTERN but convert the waveform to
% unsigned short integers so the bytes can be swapped.
% Note: Can't swap the bytes of signed short integers
    in MATLAB.
waveform = uint16(mod(65536 + waveform,65536)); %
% If on a PC swap the bytes to Big Endian
if strcmp( computer, 'PCWIN')
waveform = bitor(bitshift(waveform,-8),
    bitshift(waveform,8));
end
% Save the data to a file
% Note: The waveform is saved as unsigned short
    integers. However,
% the acual bit pattern is that of signed short
    integers and
% that is how the Agilent MXG/ESG/PSG interprets them.
```

```
filename = 'C:\Temp\PSGTestFile';
[FID, message] = fopen(filename,'w');% Open a file to
   write data
if FID == -1 error('Cannot Open File'); end
fwrite(FID,waveform,'unsigned short');% write to the
   file
fclose(FID); % close the file
% 3.) Load the internal Arb format file
***********************************
% This process is just the reverse of saving the
   waveform
% Read in waveform as unsigned short integers.
% Swap the bytes as necessary
% Convert to signed integers then normalize between +-1
% De-interleave the I/Q Data
% Open the file and load the internal format data
[FID, message] = fopen(filename,'r');% Open file to
   read data
if FID == -1 error('Cannot Open File'); end
[internalWave,n] = fread(FID, 'uint16');% read the IQ
   file
fclose(FID);% close the file
internalWave = internalWave'; % Conver from column
   array to row array
% If on a PC swap the bytes back to Little Endian
if strcmp( computer, 'PCWIN' ) % Put the bytes into
   the correct order
internalWave= bitor(bitshift(internalWave,-8),bitshift
   (bitand(internalWave,255),8));
end
% convert unsigned to signed representation
internalWave = double(internalWave);
tmp = (internalWave > 32767.0) * 65536;
iqWave = (internalWave - tmp) ./ 32767; % and
   normalize the data
% De-Interleave the IQ data
IwaveIn = iqWave(1:2:n);
QwaveIn = iqWave(2:2:n);
x2=IwaveIn+j*QwaveIn;

% Creat Wave %

len_sym=512;
nsamp=4;
len=len_sym*nsamp;
```

```
clock=16e6; %VSG Clock
SymRate=clock/nsamp;
M=4;
rolloff = 0.35;
sincro=[0; 1; 2; 3; 0; 1; 2; 3; 0; 1; 2; 3; 0; 1; 2;
   3; 0; 1; 2; 3];
%Pilot message in order to synchronize the VSA
signal=randint(len_sym-length(sincro),1,M);
signal=[sincro; signal]; %Signal to modulate
signal=[signal; signal];
constellation=[1+j*1 -1+j*1 1-j*1 -1-j*1];
modsignal=genqammod(signal,constellation);
filtorder = 80; % Filter order
delay = filtorder/(nsamp*2); % Group delay (# of input
   samples)
rrcfilter = rcosine(1,nsamp,'sqrt',rolloff,delay);
wave_4qam=rcosflt(modsignal,1,nsamp,'filter',rrcfilter);
wave_4qam=wave_4qam(1+40:1:len+40);
wave_4qam=0.5.*wave_4qam/max(abs(wave_4qam));
x_vsg=wave_4qam;
%x1=x_vsg;


Res=2^10;                          %LUT Resolution
N=4096/8;                    % number of random input
Inp=0.57*((randint(1,N)-.5)*2+j*(randint(1,N)-.5)*2)/
   2^.5;
%Random bit Generator(QPSK)
Inp=(rcosflt(Inp,1, 4,'sqrt', 0.35,10))';
Inp=Inp(41:1:end-32);
%x1=Inp.';
xi=real(x_vsg);
xq=imag(x_vsg);


%%%%%%%%%%%%%%%%%%%%%%%%%%%%%%%%%%%%%%%%%%%%%%%%%%%%%%%%
% GPIB AND DAC SIGNAL FORMAT
%%%%%%%%%%%%%%%%%%%%%%%%%%%%%%%%%%%%%%%%%%%%%%%%%%%%%%%%


AMPLITUD=8190;
CENTRO=8191;
scale=0.5;
scaleint = round(8192*scale)-1;
mx = max([max(abs(xi)) max(abs(xq))]);
xi_escalada=round(xi./mx.*scaleint+CENTRO);
xq_escalada=round(xq./mx.*scaleint+CENTRO);
%x= xi_escalada + j*xq_es
```

```
clear buffer1;
clear buffer2;
buffer1=dec2hex(xi_escalada,4);
buffer2(:,1)=buffer1(:,3);
buffer2(:,2)=buffer1(:,4);
buffer2(:,3)=buffer1(:,1);
buffer2(:,4)=buffer1(:,2);
xi_gpib=hex2dec(buffer2);
clear buffer1;
clear buffer2;
buffer1=dec2hex(xq_escalada,4);
buffer2(:,1)=buffer1(:,3);
buffer2(:,2)=buffer1(:,4);
buffer2(:,3)=buffer1(:,1);
buffer2(:,4)=buffer1(:,2);
xq_gpib=hex2dec(buffer2);
x_gpib=xi_gpib+i*xq_gpib;
x=x_gpib;

data=dlmread('IQCDMA1.txt');
I=data(:,1);
Q=data(:,2);
I=I.*0.1;
Q=Q.*0.1;
%data=dlmread('IQ1.txt');
%I1=data(:,1);
%Q1=data(:,2);
x_vsg=I+j*Q;
%x1=x_vsg;
%y_vsg=I1+j*Q1;

% Creating VSA object
%hVSA = actxserver('AgtVsaVector.Application');

%% Configuring VSA parameters %%
%hMeasurement = get(hVSA,'Measurement');
%hFrequency = get(hMeasurement,'Frequency');
%set(hFrequency,'Center',2.4e9);
%set(hFrequency,'Span',20e6);
%nsamp=4;
%ResultL=500;
%set(hMeasurement,'DemodConfig',2);
%hDemod = get(hMeasurement,'DigDemod');
%set(hDemod,'FilterAlpha',0.35); %Alpha cosine
%set(hDemod,'Format',4); %QPSK
```

```
%set(hDemod,'MeasFilter',0); %Root Raised Cosine
%set(hDemod,'RefFilter',2); %Raised Cosine
%set(hDemod,'ResultLen',ResultL); %Result length
%set(hDemod,'PointsPerSymbol',nsamp); %Points per
   symbol
%set(hDemod,'SyncSearch',1); %SyncSearch
%set(hDemod,'SyncPattern','000110110001101100011011000
   1101100011011');
%SyncPattern or pilot message
%clock=16e6; %VSG Clock
%SymRate=clock/nsamp;
%set(hDemod,'SymbolRate',SymRate); %Symbol Rate
%hDisplay = get(hVSA,'Display');
%hTraces = get(hDisplay,'Traces');
%hTrace1=get(hTraces,'Item',1);
%hTrace2=get(hTraces,'Item',2);
%hTrace3=get(hTraces,'Item',3);
%hTrace4=get(hTraces,'Item',4);
%hTrace5=get(hTraces,'Item',5);
%hTrace6=get(hTraces,'Item',6);
%set(hTrace1,'Format','vsaTrcFmtVectorIQ');
%set(hTrace1,'DataName','IQ Meas Time1');
%set(hTrace1,'Active',1);
%set(hMeasurement,'Continuous',1);
%invoke(hMeasurement,'Start');

%figure;
%plot(x_vsg(1:1024));
%figure;
%plot(y_vsg);
%IQData=x_vsg.';

IQData=x1.';
agt_sendcommand(io,'SOUR:RAD:ARB:STAT Off')
%[status, status_description] = agt_sendcommand(io,
   ':source:rad:arb:wav "seq:wave1"');
[status, status_description] = agt_waveformload(io,
   IQData,'wave1', 400000)
agt_sendcommand(io,'SOUR:RAD:ARB:STAT On')
```

B.6 DSI-SLM Scheme

```
%%%% Peak-to-Average Power Ratio Reduction %%%%
%%%%%%%%%%%%%%%%%% DSI-SLM Scheme %%%%%%%%%%%%%%%%%%%%%%
```

```
%close all;
clear all; clc;

%% Initilization
K = 256;
%DataLent=K-55;
DumLent=55;          % length of dummy signal
DataLent=K-DumLent;
U = 16;                  % number of subblocks (candidate
                               signal)
samplerate= 1;       % sample rate
QPSK_Set  = [1 -1 j -j];        % modulation seqeunce
Phase_Set = [1 -1 j -j ];       % phase sequecne
MAX_SYMBOLS  = 5e4;             % sample numbers
iteration=8;         % iteration number of DSI algorithm
PAPR_Orignal = zeros(1,MAX_SYMBOLS);  % vector
                                         initialization
PAPR_SLM_2    = zeros(1,MAX_SYMBOLS);  % vector
                                         initialization
PAPR_SLM_4    = zeros(1,MAX_SYMBOLS);
PAPR_SLM_8    = zeros(1,MAX_SYMBOLS);
PAPR_SLM_16   = zeros(1,MAX_SYMBOLS);
PAPR_DSI_SLM_2 = zeros(1,iteration);
PAPR_DSI_SLM_4 = zeros(1,iteration);
PAPR_DSI_SLM_8 = zeros(1,iteration);
PAPR_DSI_SLM_16 = zeros(1,iteration);
X       = zeros(U,K);
Index  = zeros(U,K);
%DumLent   = 55;
Gap = 55-DumLent;
GuardSig = zeros(1,Gap);
PAPRTH   = 5;                            % PAPR threshold

%%      DSI-SLM Loop

for nSymbol = 1:MAX_SYMBOLS

%Generating the dummy Sequence
DumSig = randint(1,DumLent,length(QPSK_Set))+1;
   % random dummy signal generation
%DumSig = zeros(1,55);
%DumSig = ones(1,55);
%Generation of Signal
Index1_201  = randint(1,DataLent,length(QPSK_Set))+1;
   % Input signal generation
```

```
%Pert = zeros(1,55);
Pert = zeros(1,DumLent);
Index(1,:)   = [Index1_201,Pert];        % addition of
   zeros
Index(2:U,:) = randint(U-1,K,length(Phase_Set))+1;
   % candidate signal generation (before
   multipication)
PreX  = QPSK_Set(Index1_201);   % modualtion of input
   signal
ModedDumSig = QPSK_Set(DumSig); % modualtion of dummy
   signal

%Add Dummy Sequence into the signal
%X(1,:)       = [PreX,GuardSig,ModedDumSig];
%X(1,:)       = [PreX,DumSig];
X(1,:)        = [PreX,ModedDumSig];      % multiplexing
                     input and dummy
Phase_Rot= Phase_Set(Index(2:U,:));   % random
   sequence generation
X1     = repmat(X(1,:),U-1,1);  % replicate the input
             signal
X(2:U,:) = X1.*Phase_Rot;% multiplying input signal
             with phase sequence
Xsam   = UtraSample(X,samplerate);     % oversampling
             factor by samplerate
x      = ifft(Xsam,K*samplerate,2);    % IFFT of the
             signal
Signal_Power = abs(x.^2); % find the signal power
Peak_Power   = max(Signal_Power,[],2);% find the
   maximum of the signal power
Mean_Power   = mean(Signal_Power,2);   % find the
   average of signal power
PAPR_temp    = 10*log10(Peak_Power./Mean_Power);
   % caluclate the log of the peak to average
PAPR_Orignal(nSymbol) = PAPR_temp(1);
PAPR_SLM_2(nSymbol) = min(PAPR_temp(1:2,:));
   % find the minimum PAPR of the first 2 candiate
   signal
PAPR_SLM_4(nSymbol) = min(PAPR_temp(1:4,:));
   % find the minimum PAPR of the first 4 candiate
   signal
PAPR_SLM_8(nSymbol) = min(PAPR_temp(1:8,:));
   % find the minimum PAPR of the first 8 candiate
   signal
```

```
PAPR_SLM_16(nSymbol)   = min(PAPR_temp(1:16,:));
   % find the minimum PAPR of the 16 candidate signal
minPAPR=10;


%% DSI LOOP
 for iter = 1:iteration
 if  PAPR_SLM_8(nSymbol)>PAPRTH % PAPR comparison with
    threshold
 DumSig = randint(1,DumLent,length(QPSK_Set))+1;
    % random dummy signal generation
 Index1_201  = Index1_201;
 %Pert = zeros(1,55);
 Pert = zeros(1,DumLent);
 Index(1,:)  = [Index1_201,Pert];     % adding zeros
    to the input signal
 Index(2:U,:)= Index(2:U,:);   % replicate the input
    signal
 PreX = QPSK_Set(Index1_201);   % Modulation of input
    signal
 ModedDumSig = QPSK_Set(DumSig);      % Modulation of
    dummy signal
 X(1,:)       = [PreX,ModedDumSig];    % multiplexing
    the input signal with dummy
 Phase_Rot= Phase_Set(Index(2:U,:));% random phase
    sequence generation
 X1    = repmat(X(1,:),U-1,1);   % replicate the input
    signal with dummy
 X(2:U,:) = X1.*Phase_Rot;       % multipication of
    input signal and phase sequence
 Xsam  = UtraSample(X,samplerate);    % oversampling
    factor by sample rate
 x     = ifft(Xsam,K*samplerate,2);    % IFFT of the
    candidate signals
 Signal_Power= abs(x.^2);% find the signal power
 Peak_Power  = max(Signal_Power,[],2);       % find the
    maximum of the signal power
 Mean_Power  = mean(Signal_Power,2);  % find the
    average of the signal power
 PAPR_temp    = 10*log10(Peak_Power./Mean_Power);
    % calculate the log of Peak to average power ratio
 PAPR_DSI_SLM_2 = min(PAPR_temp(1:2,:));
    % determine PAPR of DSI-SLM when M=2

 if PAPR_DSI_SLM_2<minPAPR      % comparison of PAPR
```

```
minPAPR=PAPR_DSI_SLM_2;
PAPR_SLM_2(nSymbol)=PAPR_DSI_SLM_2;
end
 %PAPR_DSI_SLM_4  = min(PAPR_temp(1:4,:));
 % determine PAPR of DSI-SLM when M=4
 %if PAPR_DSI_SLM_4<minPAPR     % comparison of PAPR
 %minPAPR=PAPR_DSI_SLM_4;
 %PAPR_SLM_4(nSymbol)=PAPR_DSI_SLM_4;
 %end


        end
      end
end

%% Calculating the CCDF
Nsym=MAX_SYMBOLS;
[n x]   = hist(PAPR_Orignal,[5:0.5:13]);
n=fliplr(cumsum(fliplr(n)));
[n1 x1] = hist(PAPR_SLM_2,[5:0.5:13]);
n1=fliplr(cumsum(fliplr(n1)));
[n2 x2] = hist(PAPR_SLM_4,[5:0.5:13]);
n2=fliplr(cumsum(fliplr(n2)));
[n3 x3] = hist(PAPR_SLM_8,[5:0.5:13]);
n3=fliplr(cumsum(fliplr(n3)));
[n4 x4] = hist(PAPR_SLM_16,[5:0.5:13]);
n4=fliplr(cumsum(fliplr(n4)));

%% Plot the CCDF
figure;
%semilogy(x,n/Nsym,x2,n2/Nsym);
semilogy(x,n/Nsym,'*-b',x1,n1/Nsym,'--r',x2,n2/Nsym,
    '--c',x3,n3/Nsym,'--k',x4,n4/Nsym,'--m')
axis ([5,13,0.0001,1]);
legend('Orignal','U=2','U=4','U=8','U=16')
title('DSISLM VS CSLM)
xlabel('PAPR0 [dB]');
ylabel('Pr[PAPR>PAPR0]');
hold on;
grid on
```

B.7 OPS-DSI Scheme

```
%%%% Peak-to-Average Power Ratio Reduction  %%%%
%%%%%%%%%%%%%%%%%  OPS-DSI Scheme %%%%%%%%%%%%%
```

```
%Close all;
clear all; clc;
%% Initilization
K = 256;
L = 55;                        %Length of dummy sequence
DataLent = K-L;
samplerate= 1;                 %Sample rate
QPSK_Set  = [1 -1 j -j];       %modulation sequence
Phase_Set = [1 -1 j -j ];      %phase sequence
MAX_SYMBOLS = 1e4;             %sample numbers
iter_OPS = 3;                  %iteration number of OPS
                                       algorithm
iter1 = 10;  %iteration number of DSI loop
U = 16;
X     = zeros(1,K-L);    % vector initialization
Index = zeros(U,K-L);    % vector initialization
DumSig = zeros(iter1,L);  % vector initialization
PAPRTH = 7;   % PAPR threshold
PAPR_Orignal = zeros(1,MAX_SYMBOLS);  % vector
   initialization
PAPR_OPS_DSI = zeros(1,MAX_SYMBOLS);  % vector
   initialization
PAPR_OPS     = zeros(1,iter_OPS);     % vector
   initialization
PAPR_DSI = zeros(1,iter1);            % vector
   initialization
Index(2:U,:) = randint(U-1,K-L,length(Phase_Set))+1;
   %random input signal generation
DumSig(1:iter1,:) = randint(iter1,L,length(QPSK_Set))+1;
   %random dummy signal generation


%% OPS-DSI Loop

for nSymbol = 1:MAX_SYMBOLS
    X  = QPSK_Set(randint(1,DataLent,length(QPSK_
          Set))+1); % modualtion of input signal

  for iter2 = 1:iter_OPS


    Phase_Rot = Phase_Set(Index(2:U,:));
       % Random phase sequence generation
    X2 = X.*Phase_Rot(iter2,:);
       % multiplying input signal with phase sequence
```

```
    for iter1 = 1:iter1

        ModedDumSig = QPSK_Set(DumSig(1:iter1,:));
            % modulation of dummy signal
        X3    = [X2,ModedDumSig(iter1,:)];
            % multiplexing the signal and dummy
        Xsam = UltraSample(X3,samplerate);
            % oversampling factor by samplerate
        x = ifft(Xsam,K*samplerate);     % IFFT of the
            signal
        Signal_Power = abs(x.^2);% find the signal power
        Peak_Power    = max(Signal_Power,[],2);
            % find the maximum of the signal power
        Mean_Power    = mean(Signal_Power,2);  % find the
            average of signal power
        PAPR_DSI(iter1)  = 10*log10(Peak_Power./Mean_
            Power); % calculate the log of PAPR
    end

  if min(PAPR_DSI) < PAPRTH % comparison of PAPR
   PAPR_OPS_DSI(nSymbol)= min(PAPR_DSI);

   end
  PAPR_OPS(iter2)= min(PAPR_DSI);
end

  PAPR_OPS_DSI(nSymbol)= min(PAPR_OPS);
  x = ifft(X,K*samplerate,2);     % IFFT of the signal
  Signal_Power = abs(x.^2);
  Peak_Power    = max(Signal_Power,[],2);     % find the
     maximum of the signal power
  Mean_Power    = mean(Signal_Power,2);
  PAPR_Orignal(nSymbol)  = 10*log10(Peak_Power./Mean_
     Power);

end

%%   Calculating the CCDF
Nsym=MAX_SYMBOLS;
[n x]    = hist(PAPR_Orignal,[5:0.5:13]);
n = fliplr(cumsum(fliplr(n)));
[n1 x1] = hist(PAPR_OPS_DSI,[5:0.5:13]);
n1 = fliplr(cumsum(fliplr(n1)));

%% Plot the CCDF
figure;
%semilogy(x,n/Nsym,x2,n2/Nsym);
```

```
semilogy(x,n/Nsym,'-b',x1,n1/Nsym,'--r')
%axis ([5,13,0.01,1]);
legend('Orignal','iter-OPS = 8, & iter-DSI = 10')
title('OPS-DSI SCHEME FOR PAPR REDUCTION')
xlabel('PAPR0 [dB]');
ylabel('CCDF (Pr(PAPR>PAPR0)');
hold on;
grid on
```

B.8 Power Amplifier

```
%%%% Power Amplifier Model with Memory Effects  %%%%%

function Out=amplifiermodel(Inp)

data=dlmread('Test Bed.txt');   % Read the imported
    samples
Pin=data(:,1);                  % Input Power in dBm
Pindbm=Pin;
Pout=data(:,2);                 % Output Power in dBm
Phase=data(:,3);                % Output Phase in deg
%Phase=Phase.*pi./180;
%Inp=Pin;
Pout=sqrt(10.^((Pout./10)-3));  % Converting the Output
    Power to Watt
Pin=sqrt(10.^((Pin./10)-3));    % Converting the Input
    Power to Watt
%Gain=(Pout-Pin);
%Gain1=20.*log10((Pout./Pin));
%Gmax=max(Gain);
Pinmax=max(Pin);                % Identifying the
    maximum input power
Poutmax=max(Pout);              % Identifying the
    maximum output power
Pin=(Pin./Pinmax);              % Normalizing the input
    power
Pout=(Pout./Pinmax);            % Normalizing the
    output power
Gain1=((Pout./Pin));            % find the power
    amplifier gain

%Gain=(Pout./Pin);
%Inp=4.*Inp;
u=abs(Inp);                     % absolute value of
    Input signal
```

```
AMAMco=polyfit(Pin,Pout,12);    % find the coefficients
   of the polynomial AMAMco of degree 12
AMPMco=polyfit(Pin,Phase,12);   % find the coefficients
   of the polynomial AMPMco of degree 12

AMAM=polyval(AMAMco,abs(Inp));  % returns the value of
   polynomial AMAMco of degree 12 evaluated at abs(Inp)
%AMAM=AMAM./max(AMAM);
AMPM=polyval(AMPMco,u);         % returns the value of
   polynomial AMPMco of degree 12 evaluated at abs(Inp)
Phi=unwrap(AMPM+angle(Inp));    % corrects the radian
   phase angles in a vector AMPM+angle(Inp)
Out=AMAM.*exp(j*Phi);    % Output signal generation
```

B.9 Parameter Initialization of OPS-DSI Implementation

```
%%%%%%%%%% Parameter initialization %%%%%%%%%%%%%

clear;
clc;
Z=100;                      % Number of Samples
M=16;                       % Number of subblocks
P=16;                       % Number of OPS iteration
N=256;                      %IFFT length
NN=N*M;
I=10;                       % Number of DSI iteration
K=N*Z;
PAPR_Orignal = zeros(1,Z);    % Vector initialization
sync= zeros(N,1);           % Vector initialization
sync(1,1) = 1;              % synchronization
QPSK_Set  = [1 -1 j -j];    % modulation sequence
Index = randint(1,K,4)+1;   % phase sequence index
   initialization
XX = QPSK_Set(Index(1,:));  % modulation of input signal
XX = XX.';                  % Input Signal
XX = repmat(XX,M,1);        % replicate the input signal
XX1= reshape(XX,N,M*Z*S);       % Dimension conversion
   for FPGA
O11= zeros(NN,Z);               % Vector initialization

for n=1:Z
  O1=XX1(:,n);
  O11(:,n)=repmat(O1,M,1);   % Dimension conversion
     for FPGA
end
```

```
XX2=reshape(O11,[],1);          % Dimension conversion
   for FPGA
Phase_Set = [1 -1 j -j];        % Phase sequence
Index1 = randint(1,NN,4)+1;     % Dimension conversion
   for FPGA
Phase = Phase_Set(Index1(1,:)).';   % phase sequence
   generation
Phase = repmat(Phase,Z,1);      % Dimension conversion
   for FPGA
%Inp=XX2.*Phase;
XX3=reshape(XX,[],Z);     % Dimension conversion for
   FPGA

for n=1:Z
 XX4=XX3(:,n);
 x=ifft(XX4,N);     % IFFT of original OFDM Signal
 Signal_Power = abs(x.^2);       % find the signal power
 Peak_Power   = max(Signal_Power); % find the maximum
    of signal power
 Mean_Power   = mean(Signal_Power); % find average of
    signal power
 PAPR_Orignal(n)= 10*log10(Peak_Power./Mean_Power);
    % find PAPR of original OFDM signal
end

Nsym=L;
[n x] = hist(PAPR_Orignal,[4:0.5:13]);
n=fliplr(cumsum(fliplr(n)));
%x=[1+0.14:0.14:10];
figure;
semilogy(x,(n)/Nsym,'LineWidth',2)
axis([4,13,0.01,1])
```

B.10 Program for Experimental Measurement

```
clear;
io = agt_newconnection('gpib',0,20);
(status, status_description,query_result) = agt_
   query(io,'*idn?');
if (status < 0) return; end
agt_sendcommand(io,'SOUR:RAD:ARB:STAT Off')
agt_sendcommand (io, 'FREQuency 2400000000')
agt_sendcommand (io, 'POWer -7')
```

```
( status, status_description ) = agt_
   sendcommand( io, 'OUTPut:STATe ON' );
( status, status_description ) = agt_
   sendcommand( io, 'POW:ALC:STAT Off');


PA_Model=11;
% INPUT SIGNAL
load X-5.mat;
load F.mat;
x1=F.*x1;

% and a sine waveform in Q.
%
points = 1000; % Number of points in the waveform
cycles = 101; % Determines the frequency offset from
   the carrier
phaseInc = 2*pi*cycles/points;
phase = phaseInc * (0:points-1);
Iwave = cos(phase);
Qwave = sin(phase);
% 2.) Save waveform in internal format **************
% Convert the I and Q data into the internal arb format
% The internal arb format is a single waveform
   containing interleaved IQ
% data. The I/Q data is signed short integers
   (16 bits).
% The data has values scaled between +-32767 where
% DAC Value Description
% 32767 Maximum positive value of the DAC
% 0 Zero out of the DAC
% -32767 Maximum negative value of the DAC
% The internal arb expects the data bytes to be in Big
   Endian format.
% This is opposite of how short integers are saved on
   a PC (Little Endian).
% For this reason the data bytes are swapped before
   being saved.
% Interleave the IQ data
waveform(1:2:2*points) = Iwave;
waveform(2:2:2*points) = Qwave;
%(Iwave;Qwave);
% Normalize the data between +-1
waveform = waveform / max(abs(waveform)); % Watch out
   for divide by zero.
% Scale to use full range of the DAC
```

```
waveform = round(waveform * 32767); % Data is now
    effectively signed short integer values
% waveform = round(waveform * (32767 /
    max(abs(waveform)))); % More efficient than
    previous twosteps!
% PRESERVE THE BIT PATTERN but convert the waveform to
% unsigned short integers so the bytes can be swapped.
% Note: Can't swap the bytes of signed short integers
    in MATLAB.
Waveform = uint16(mod(65536 + waveform,65536)); %
% If on a PC swap the bytes to Big Endian
if strcmp( computer, 'PCWIN')
waveform = bitor(bitshift(waveform,-8),
    bitshift(waveform,8));
end
% Save the data to a file
% Note: The waveform is saved as unsigned short
    integers. However,
% the acual bit pattern is that of signed short
    integers and
% that is how the Agilent MXG/ESG/PSG interprets them.
Filename = 'C:\Temp\PSGTestFile';
(FID, message) = fopen(filename,'w');% Open a file to
    write data
if FID == -1 error('Cannot Open File'); end
fwrite(FID,waveform,'unsigned short');% write to the
    file
fclose(FID); % close the file
% 3.) Load the internal Arb format file **************
% This process is just the reverse of saving the
    waveform
% Read in waveform as unsigned short integers.
% Swap the bytes as necessary
% Convert to signed integers then normalize between +-1
% De-interleave the I/Q Data
% Open the file and load the internal format data
(FID, message) = fopen(filename,'r');% Open file to
    read data
if FID == -1 error('Cannot Open File'); end
(internalWave,n) = fread(FID, 'uint16');% read the IQ
    file
fclose(FID);% close the file
internalWave = internalWave'; % Convert from column
    array to row array
% If on a PC swap the bytes back to Little Endian
```

```
if strcmp( computer, 'PCWIN' ) % Put the bytes into
  the correct order
internalWave= bitor(bitshift(internalWave,-8),bitshift
  (bitand(internalWave,255),8));
end
% convert unsigned to signed representation
internalWave = double(internalWave);
tmp = (internalWave > 32767.0) * 65536;
iqWave = (internalWave - tmp) ./ 32767; % and
  normalize the data
% De-Interleave the IQ data
IwaveIn = iqWave(1:2:n);
QwaveIn = iqWave(2:2:n);
x2=IwaveIn+j*QwaveIn;


% Create Wave %

len_sym=512;
nsamp=4;
len=len_sym*nsamp;
clock=16e6; %VSG Clock
SymRate=clock/nsamp;
M=4;
rolloff = 0.35;
sincro=(0; 1; 2; 3; 0; 1; 2; 3; 0; 1; 2; 3; 0; 1; 2;
  3; 0; 1; 2; 3);
%Pilot message in order to synchronize the VSA
signal=randint(len_sym-length(sincro),1,M);
signal=(sincro; signal); %Signal to modulate
signal=(signal; signal);
constellation=(1+j*1 -1+j*1 1-j*1 -1-j*1);
modsignal=genqammod(signal,constellation);
filtorder = 80; % Filter order
delay = filtorder/(nsamp*2); % Group delay (# of input
  samples)
rrcfilter = rcosine(1,nsamp,'sqrt',rolloff,delay);
wave_4qam=rcosflt(modsignal,1,nsamp,'filter',rrcfilter);
wave_4qam=wave_4qam(1+40:1:len+40);
wave_4qam=0.5.*wave_4qam/max(abs(wave_4qam));
x_vsg=wave_4qam;
%x1=x_vsg;


Res=2^10;                    %LUT Resolution
N=4096/8;                    % number of random input
```

```
Inp=0.57*((randint(1,N)-.5)*2+j*(randint(1,N)-.5)*2)/
   2^.5;     %Random bir Generator(QPSK)
Inp=(rcosflt(Inp,1, 4,'sqrt', 0.35,10))';
Inp=Inp(41:1:end-32);
%x1=Inp.';
xi=real(x_vsg);
xq=imag(x_vsg);


%%%%%%%%%%%%%%%%%%%%%%%%%%%%%%%%%%%%%%%%%%%%%%%%%%%%%%
% GPIB AND DAC SIGNAL FORMAT
%%%%%%%%%%%%%%%%%%%%%%%%%%%%%%%%%%%%%%%%%%%%%%%%%%%%%%


AMPLITUD=8190;
CENTRO=8191;
scale=0.5;
scaleint = round(8192*scale)-1;
mx = max((max(abs(xi)) max(abs(xq))));
xi_escalada=round(xi./mx.*scaleint+CENTRO);
xq_escalada=round(xq./mx.*scaleint+CENTRO);
%x= xi_escalada + j*xq_es
clear buffer1;
clear buffer2;
buffer1=dec2hex(xi_escalada,4);
buffer2(:,1)=buffer1(:,3);
buffer2(:,2)=buffer1(:,4);
buffer2(:,3)=buffer1(:,1);
buffer2(:,4)=buffer1(:,2);
xi_gpib=hex2dec(buffer2);
clear buffer1;
clear buffer2;
buffer1=dec2hex(xq_escalada,4);
buffer2(:,1)=buffer1(:,3);
buffer2(:,2)=buffer1(:,4);
buffer2(:,3)=buffer1(:,1);
buffer2(:,4)=buffer1(:,2);
xq_gpib=hex2dec(buffer2);
x_gpib=xi_gpib+i*xq_gpib;
x=x_gpib;

 data=dlmread('IQCDMA1.txt');
I=data(:,1);
Q=data(:,2);
I=I.*0.1;
Q=Q.*0.1;
%data=dlmread('IQ1.txt');
```

```
%I1=data(:,1);
%Q1=data(:,2);
x_vsg=I+j*Q;
%x1=x_vsg;
%y_vsg=I1+j*Q1;

% Creating VSA object
%hVSA = actxserver('AgtVsaVector.Application');

%% Configuring VSA parameters %%
%hMeasurement = get(hVSA,'Measurement');
%hFrequency = get(hMeasurement,'Frequency');
%set(hFrequency,'Center',2.4e9);
%set(hFrequency,'Span',20e6);
%nsamp=4;
%ResultL=500;
%set(hMeasurement,'DemodConfig',2);
%hDemod = get(hMeasurement,'DigDemod');
%set(hDemod,'FilterAlpha',0.35); %Alpha cosine
%set(hDemod,'Format',4); %QPSK
%set(hDemod,'MeasFilter',0); %Root Raised Cosine
%set(hDemod,'RefFilter',2); %Raised Cosine
%set(hDemod,'ResultLen',ResultL); %Result length
%set(hDemod,'PointsPerSymbol',nsamp); %Points per
    symbol
%set(hDemod,'SyncSearch',1); %SyncSearch
%set(hDemod,'SyncPatt
    ern','00011011000110110001101100011011000011011');
%SyncPattern or pilot message
%clock=16e6; %VSG Clock
%SymRate=clock/nsamp;
%set(hDemod,'SymbolRate',SymRate); %Symbol Rate
%hDisplay = get(hVSA,'Display');
%hTraces = get(hDisplay,'Traces');
%hTrace1=get(hTraces,'Item',1);
%hTrace2=get(hTraces,'Item',2);
%hTrace3=get(hTraces,'Item',3);
%hTrace4=get(hTraces,'Item',4);
%hTrace5=get(hTraces,'Item',5);
%hTrace6=get(hTraces,'Item',6);
%set(hTrace1,'Format','vsaTrcFmtVectorIQ');
%set(hTrace1,'DataName','IQ Meas Time1');
%set(hTrace1,'Active',1);
%set(hMeasurement,'Continuous',1);
%invoke(hMeasurement,'Start');
```

```
IQData=x1.';
agt_sendcommand(io,'SOUR:RAD:ARB:STAT Off')
%(status, status_description) = agt_sendcommand(io,
   ':source:rad:arb:wav "seq:wave1"');
(status, status_description) = agt_waveformload(io,
   IQData,'wave1', 400000)
agt_sendcommand(io,'SOUR:RAD:ARB:STAT On')
```

Index